NORTHERN LIGHT AND NORTHERN TIMES

Swedish Leadership in the Foundation
of Biological Rhythms Research

NORTHERN LIGHT AND NORTHERN TIMES

Swedish Leadership in the Foundation of Biological Rhythms Research

Jole Shackelford

American Philosophical Society
Philadelphia • 2013

Transactions of the
American Philosophical Society
Held at Philadelphia
For Promoting Useful Knowledge
Volume 103, Part 2

ISBN: 978-1-60618-032-7
US ISSN: 0065-9746

Library of Congress Cataloguing-in-Publication Data

Shackelford, Jole, author.
  Northern light and northern times : Swedish leadership in the foundation of
biological rhythms research / Jole Shackelford.
     pages cm. — (Transactions of the American Philosophical Society, ISSN
0065-9746 ; volume 103, part 2)
  Includes bibliographical references and index.
  ISBN 978-1-60618-032-7 (alk. paper)
  1. Biological rhythms—Research—Sweden—History—20th century. I. Title.
QH527.S43 2013
571.7'709485—dc23
                                    2013042644

Contents

Acknowledgments

The research behind this work draws on general reading in the history of biological rhythms research undertaken in part with the support of NSF grant SES-0958974, but also on research conducted on site in Ronneby, Sweden, and the National Archives and Royal Library in Stockholm. I am grateful to Hans Göran Eriksson for his hospitality, which made my visit to Ronneby Brunn possible; to archivist Anders Karlsson at the Ronneby stadsarkiv; and to Björg Elisabeth Hammam Lie, Librarian at the Royal Institute of Technology (KTH), who helped me navigate the research venues of Stockholm and made my stay there both pleasant and productive. My thanks, too, to Frankie Shackelford for commenting on the manuscript.

Preface

The consequences of jet lag, shift work, sleep disorders, and seasonal affective disorder are common knowledge among the public in industrialized nations and, with globalization, this knowledge is increasingly disseminated worldwide. These phenomena involve one or more aspects of the internal rhythmic structure of human physiology, commonly called biological rhythms. There is an abundance of popular literature on human biological rhythms— how to understand them and use them to maximize health, productivity, sexual performance, weight loss, and any number of other factors associated with modern living. Moreover, there is an even greater abundance of scientific literature on biological rhythms, from the rhythmic behaviors of key chemical reactions in single-cell plants and animals to the developmental characteristics of molds and fungi, to the complex timing of endocrine systems in mammals. There are biological rhythms in physiological systems, behavioral patterns, and population fluctuations. Biological rhythms are implicated in the seasonal maturation and blooming of plants, the communication among bees regarding the distances and directions from their hive to sources of food, and the celestial navigation of birds migrating across continents. These biological rhythms may have characteristic periods of mere seconds to years, but the great majority of rhythms that have been studied are daily rhythms, called *circadian rhythms* since 1959, or tidal (about half-daily) rhythms. Their study is part of a larger field of biological research called *chronobiology*.

Chronobiology comprises the study of all temporal aspects of biological organization, including biological rhythmicity, but also developmental timing in embryology, population variations with time, the duration of reproductive cycles (often called "the biological clock"), senescence, and death. The term *chronobiology* gained currency in the 1970s, but is still not entirely accepted within biological rhythms studies, as is evidenced by the various titles of learned societies and journals within the field. Its growing acceptance among medical scientists in particular is evident in the literature. The historical development of chronobiology is rooted in biological rhythms research, which began in the eighteenth and nineteenth centuries, but did not become internationally organized with well-articulated principles and methods until the twentieth century. In the early decades of the twentieth century, biologists were discovering and investigating the rhythmic behaviors of plants in research laboratories and animals in the field in the United States, Canada, India, throughout Europe, and most everywhere that Western science reached, but the chief research on the nature and causes of these rhythms was being done in northern Europe. Given the prominence of German laboratory science in the nineteenth century, it follows that German biologists were at the forefront of biological rhythms research in the early twentieth century. The formative years of the first international society of biological rhythms researchers, the *Societas pro studio rythmi biologici* (Society for the Study of Biological Rhythm) exhibit the influence of German scientists, but surprisingly also the importance of Dutch and, in particular, Swedish researchers, many of whom had studied in German laboratories. Owing to various intellectual and political circumstances, which will be explored in this monograph, Swedish biologists and physicians were crucial to the early development of the international Society and, thus, the fostering of the new academic research field of chronobiology.

The history of biological rhythms research and chronobiology more broadly has, despite the abundance of popular and scientific publications devoted to the science, received very little attention from academic historians of science, technology, and medicine. This may be partly a consequence of the prominence of heredity and evolution in twentieth-century biology and of the emphasis placed on molecular biology after the discovery of DNA and the genetic revolution that has followed. Historical narratives of the genealogy of biological rhythms research can be found in various scientific papers, but these are by nature partisan and reflect scientists' perspectives and priorities, not the views and methods of academic historians. Moreover, even within these narratives, the work of the early Scandinavian researchers is diminished relative to their salience in the first decades of the twentieth century, owing to the dominance of the German, French, and English languages in the sciences and the roles of a small cohort of German, English, French, and American scientists in bringing the field to prominence in the 1950s and 1960s. The history of science, too, has been dominated by these language groups, having a decidedly Anglo-American perspective during the twentieth century, even as it reflected a strong presence of German, French, and to a lesser extent Swedish and Italian national literatures. For all these various reasons, this initial foray into the subject of the history of biological rhythms research examines the early professional organization of the field, in which the efforts and scientific research of the early participants, many of them Swedes, comes to light.

1

Disciplining Biological Rhythms

Living organisms, from single-cell prokaryotes to humans, exhibit a variety of rhythmic behaviors. These occur at the cellular and subcellular levels and, in complex organisms, coordinate intricate biochemical systems and direct the activities of individuals on tidal, daily, weekly, monthly, and annual periodicities. The scientific investigation of the rhythmic behaviors of living organisms—of their internal physiological patterns as well as the temporality of their interactions with the ecosystem—is by its very nature multidisciplinary and interdisciplinary, as is evident in its historical development. The modern study of biological rhythms began with isolated experiments that are documented in dissertations and research papers written in the decades around the turn of the twentieth century, work that was undertaken by scattered researchers within the existing disciplinary frames of zoology, meteorology, botany, cell biology, chemistry, physiology, pathology, epidemiology, and other scientific fields. Rhythmic behaviors were manifest in all these disciplinary contexts, but the existing disciplinary boundaries also acted to isolate researchers from the cross-disciplinary exchanges of ideas and cooperation that ultimately led to the development of biological rhythms research as a coherent science. The development of such cross-disciplinary networks was a crucial component of the institutionalization of the field.[1] The first pivotal moment in the professional development of biological rhythms studies was a meeting of German, Dutch, and Scandinavian biologists outside a small town in southern Sweden in 1937 that led to the formation of the *Societas pro studio rythmi biologici*, known more widely for many years by its German name, the Internationale Gesellschaft für Biologische Rhythmusforschung. That this meeting was organized in Sweden and the resultant Society sponsored and dominated by Swedes in its formative decades can be explained by a tradition of Swedish research on phenomena related to solar and lunar periodicities in the Humboldtian tradition of cosmic physics, the Karolinska Institute's early funding of diabetes research, and Sweden's position as a neutral power during and after the Second World War. The intellectual development of rhythms studies leading up to this meeting, its social and organizational aspects, and its consequences for institutionalizing the field are the focus of this book.

The earliest scientific investigations of rhythmic behaviors of plants and animals, of cells, systems, and communities were carried out by widely scattered scholars working generally in isolation from other rhythms researchers in backyard laboratories in Illinois or marine biological stations in Cuba and

---

[1] Erwin Bünning, "Fifty Years of Research in the Wake of Wilhelm Pfeffer," *Annual Review of Plant Physiology* 28 (1977): 1–22, p. 15, explicitly identified disciplinary boundaries as a hindrance: "It became evident that we were dealing with a more general phenomenon.... This stimulated me already at that time to incorporate insects into my research program.... But when in 1935 I announced a paper on endogenous diurnal periodicity in animals and plants for the annual meeting of the German Botanical Society, I was asked whether I should not better restrict myself to plants, since this was a meeting of botanists."

Bermuda, or in a quiet harbor on the Isle of Man—individual projects that sometimes quite unexpectedly yielded revelations about the rhythms of life. The intellectual foundation of the field was therefore the product of individual scientists working in diverse local and national contexts. However, the creation of an international society, conceived at the outset to bring together botanists, zoologists, and clinical researchers to share their common passion for the study of rhythmic behaviors in the organic world informs us about how the science moved from the researcher's study and the laboratory into a more public space, becoming a truly collective, international activity.[2] The creation of the *Societas pro studio rythmi biologici* (the *Societas* or Society) represents the collectivization of these lines of scientific inquiry across disciplinary boundaries and the forging of a group identity that can be called a scientific field or subdiscipline, perhaps even a discipline in its own right.[3] The institutional lineage of the *Societas* still exists, but under a new name to reflect a new sense of disciplinary identity that its members sought in the second half of the twentieth century: The International Society for Chronobiology.

Alberto Cambrosio and Peter Keating used chronobiology in their 1983 article as a case study of discipline formation, drawing attention mainly to how organizational development, rhetoric, and leadership all play important roles in creating identity for a specialty as it heads toward what can be defined as disciplinary status.[4] Their study, although partly historical in nature, mainly served to elaborate a sociology of science and reflects a historical perspective that privileges the genealogical narratives of two particular approaches to biological rhythms research and the struggle to control the direction of the field's professionalization subsequent to the 1960 Cold Spring Harbor

---

[2] Model studies of the institutionalization of science in this period are Robert E. Kohler's, *From Medical Chemistry to Biochemistry: The Making of a Biomedical Discipline* (Cambridge: Cambridge University Press, 1982) and William Bechtel's, *Discovering Cell Mechanisms: The Creation of Modern Cell Biology* (Cambridge: Cambridge University Press, 2006).

[3] The distinctions among "fields," "specialties," "subdisciplines," "disciplines," etc. are imprecise, and definitions are debated within the history and sociology of science. The etymology of the term "discipline" recommends the distinction between "discipline" as "teaching domain" and "specialty" as "research domain" that is offered by Daryl E. Chubin, "The Conceptualization of Scientific Specialties," *Sociological Quarterly* 17 (1976): 448–76, p. 448. However, the distinctions that Chubin elaborates in this article are better suited to the characterization of research specialties that break off of and are subalternate to existing disciplines than they are to research specialties that develop across or between disciplines. The history of biological rhythms research suggests that it is of the latter type, which does not fit well into the discipline-formation schemata offered by Warren Hagstrom, Chubin, and their kind (cf. Ibid., p. 450).

[4] The transformation of the *Societas pro studio rythmici biologici* into the International Society for Chronobiology is the subject of Alberto Cambrosio and Peter Keating's, "The Disciplinary Stake: The Case of Chronobiology," *Social Studies of Science* 13 (1983): 323–53, which examines the sociology of discipline formation in the sciences. The present study explains the history behind this sociological process.

Symposium on Biological Clocks.[5] The nature of scientific disciplines and their formative stages, characteristics, and processes have been subject to scrutiny by numerous authors since, and current approaches recognize the importance of intellectual and epistemological developments as well as sociological determinants in the shaping of research areas, subject specialties, and disciplines.[6] I do not intend here to engage in an examination of what constitutes a discipline or whether chronobiology today can be considered a discipline, but rather to examine from a historian's perspective an early stage in its development, namely, its formal organization as an international research area with a coherent subject matter, guiding questions, methodologies, and scholarly network. The 1937 founding of the *Societas* to serve the common interests of researchers working in diverse existing disciplines and nations and to provide them a collective, international identity and platform for discipline formation lies at the center of this study, which begins with the recognition of biological rhythms as phenomena that evoke questions requiring a cross-disciplinary investigation and continues to the mid-1950s. By that time biological rhythms study was on its way to becoming international chronobiology.[7]

---

[5] Cambrosio and Keating framed the disciplinary process as a struggle between competing visions for the field held by Colin Pittendrigh and Franz Halberg and corresponded with both of these researchers while researching their article. They also drew perspectives on the history of biological rhythms research from Richie R. Ward, *The Living Clocks* (New York: Knoff, 1971), whose account privileges Colin Pittendrigh's leadership in the emergent field of biological rhythms research, and also from various articles and books by Erwin Bünning, who likewise became a strong supporter of Pittendrigh, as did Jürgen Aschoff. These three were consequently valorized as key fathers of modern biological rhythms research, against which Halberg strove to create chronobiology. Cambrosio and Keating's analysis was based on solid research, but both their sociological aims and choice of a few recent key actors to exemplify the larger historical process introduced simplifications and partisan perspectives that distort the historical narrative.

[6] Alexander Powell, Maureen A. O'Malley, Staffan Müller-Wille, Jane Calvert, and John Dupré, "Disciplinary Baptisms: A Comparison of the Naming Stories of Genetics, Molecular Biology, Genomics, and Systems Biology," *History and Philosophy of the Life Sciences* 29 (2007): 5–32, provides a useful overview of the disciplinary process as it variously applied to the four cases named in the title, but the authors' emphasis is clearly on the importance of nomenclature in conferring collective identity. Of these four, the discipline formation of genetics best fits chronobiology, owing in part to its emergence from a multidisciplinary background (p. 12). Although the sociological approaches have been criticized by K. Brad Wray, "Rethinking Scientific Specialization," *Social Studies of Science* 35 (2005): 151–64, for subordinating cognitive aspects of scientific inquiry to the dynamics of social structures (p. 154), even Chubin, "Conceptualization of Scientific Specialties," p. 470, acknowledged that a pluralistic approach is ideal: "To reconstruct the development of a specialty requires knowledge of its intellectual history plus structural (bibliographic and sociometric) analyses of its members."

[7] The creation and deployment of the branding name "chronobiology" is central to Cambrosio and Keating's analysis and fits well the criteria discussed by Powell et al., "Disciplining Baptisms," p. 8, which would place the "baptism" of the specialty field or discipline of biological rhythms research in the 1960s. But by this time the key intellectual problems, observations, and philosophical frameworks for interpretation were decades old. The logic of biological rhythms constituted a special cognitive area that was in parts common to multiple existing disciplines, but was well served by none, and long preceded the more evident social construction of the specialty as chronobiology.

2

Wilhelm Pfeffer and the Roots of Twentieth-Century Biological Rhythms Research

Chronobiology is the term now widely applied to the study of rhythmic behavior in biological organisms and their component parts as well as the deeper evolutionary significance and biomedical consequences of biological rhythms. The circadian rhythms, or those that characterize patterns of organic behavior that recur with an "about-24-hour period," are the most obvious, but there are rhythms in everything from the transcription of DNA in cells to the mechanics of organ graft rejections, which manifest periodicities from fractions of a day to weeks, months, and even years.[1] The spectrum of cyclical organic behavior is rich, presumably an evolutionary response to the advantages of cycling the storage and usage of energy and to the fitness gained by anticipating periodically changing environments.[2] When scientists first began to recognize the nature and significance of biological rhythms, to study them by intention and by systematic experimentation, and to theorize, they did not use the term *chronobiology*, but rather described them as cycles, periodicities, and increasingly as rhythms.

Recognition of organic rhythms is very old, dating back to the ancient Babylonians and, therefore, to the beginnings of our historical tradition. Hippocratic medicine is founded on the idea that the condition of the body changes with changing environmental stimuli and that health and disease are contingent on variable "airs, waters, and places" as well as on seasons, time of day, and time of life. However, the notion that rhythmic behavior is an innate characteristic of organisms rather than merely a response to rhythmic external stimuli first appeared in European science during the early-modern period, and then only gradually.[3] During the sixteenth century it was hypothesized that the timings of organic chemical behavior were internal in nature, microcosmic

---

[1] Rhythms with periods significantly shorter than 24 hours are classified as ultradian and those with periods longer than about 28 hours as infradian. The term "circadian," on which these terms are based, was publicly launched in 1959 by Franz Halberg, whose current work is on cycles with periods greater than a decade, which apparently correspond with solar activity cycles.

[2] Put briefly, birds that are bred to get up early are more successful than their later-rising fellows at finding nocturnal worms, and those worms that go to sleep before the birds get up live longer than later-active individuals, resulting in *inherent* diurnal and nocturnal habits. There has been speculation that diurnal cycles in transparent unicellular life forms may have evolved to shield radiation-sensitive molecular processes from harmful ultraviolet rays by coordinating them at night.

[3] One might think that the Hippocratic concept of critical days, which is based on the observation that the course of many illnesses is periodic after the initial onset, implies that the early Greeks understood biological rhythms. However, this was not a phenomenon recognized as innate to the human body, but rather as a property of its response to an altered balance of qualities and fluids, which in the majority of cases could be linked to the environment through what was elaborated in medieval medicine as a system of six "non-naturals." Inasmuch as the consensus of Hippocratic texts does not portray diseases as "things" but rather as improper mixtures and imbalanced conditions, the periodic behavior of diseases was not an essential property of the human being (or other afflicted organism), but an accident of pathology. On Galen's use of Hippocratic critical days, see Glen Cooper, "Galen and Astrology: A Mésalliance?" *Early Science and Medicine* 16 (2011): 120–46.

analogues of the timings evident in the celestial world. Several eighteenth-century natural philosophers observed that plants "sleep" at night and recognized that certain plants, notably the heliotropes, respond to the position of the sun and will continue to do so when apparently kept in the dark.[4] For Cartesian astronomers this was a beautiful illustration of the complexity and ingenuity of the world machine. But in the nineteenth century, scientists began to investigate rhythmic phenomena more systematically and began to realize that the rhythms discerned in various organisms showed a measure of independence from the timings of the environmental stimuli to which they were exposed. This combination of methodological refinement and theoretical development marks a watershed between the age-old recognition that organisms show microcosmic rhythmic behavior in a rhythmic macrocosm and the modern understanding that timing can be a complex physiological process with an underlying structural cause. From a philosophical point of view, identifying the causal, structural underpinning of manifest physiological properties is important, and the search for what scientists called "the biological clock" or some variant of this name became increasingly urgent in the twentieth century.

Perhaps the first study to meet these criteria was undertaken by Julien-Joseph Virey (1775–1846). Virey was a French pharmacist with medical–scientific ambitions, and he systematically studied daily and annual rhythms of human mortality and published both his data and his results in his doctoral thesis, which he defended at the University of Paris Faculty of Medicine in 1814. Virey's methodology is recognizably modern, and he speculated that biological timing was not merely the organism's response to periodic environmental stimuli, but was endogenous. For both these reasons, Virey's work is a convenient marker between the prehistory of biological rhythms research and its modern development as an experimental and exact science.[5]

---

[4] Erwin Bünning, "Opening Address: Biological Clocks," in *Biological Clocks*, Cold Spring Harbor Symposia on Quantitative Biology, v. 25 [1960] (Cold Spring Harbor, NY: The Biological Laboratory, 1961), 1–9, p. 1, identified Jean Jacques d'Ortous de Mairan, *Histoire de Académie Royale des Sciences* (Paris, 1729), as "the first to describe experiments showing that plants maintained in constant darkness and at a relatively constant temperature demonstrate diurnally periodic leaf movements in much the same way as if the plants were exposed to the normal light-dark alternation." De Mairan's observations were confirmed by Henri-Louis Duhamel du Monceau in 1758 and Johann Gottfried Zinn in 1759, which led Augustin Pyramus de Candolle, *Physiologie végétale* (Paris: Béchat-Jeune, 1832), to experiment with continuous illumination and counter-cyclical synchronization of the sensitive plant *Mimosa pudica*, revealing a shorter-than-24-hour untriggered cycle.

[5] Alain E. Reinberg, Hadas Lewy, and Michael Smolensky, "The Birth of Chronobiology: Julien Joseph Virey 1814," *Chronobiology International* 18 (2001): 173–86 point out that Virey judged that one would expect mortality statistics to be uniform if biological rhythms were not implicated in processes that are crucial for survival. To test this idea, Virey recorded the hour and date of deaths at the Val-de-Grâce hospital between May 1797 and May 1808 and concluded that vital processes in the human body are rhythmical in nature (p. 179). Because of this systematic experimentation, publication of the resultant time series, and reasoning about the endogenous nature of biological rhythms, the authors argue that "Virey deserves credit for establishing the field of chronobiology based on his insights and writings" (173–4).

Rhythmic behavior was of interest to physiologists during the middle of the nineteenth century, but it was not the rhythmicity itself that was the subject of study, but rather the cycle as a phenomenon, not its temporal characteristics. The best known example of this is perhaps Claude Bernard's recognition of the role of the liver in cycles of glycogen production. Concomitant with the development of cell biology, experimental physiologists gradually understood energetics and the cyclicity of metabolisms in plants and animals. But the actual timings of these and their nature did not attract much attention until late in the century, perhaps stimulated by Charles Darwin's interest, toward the end of his life, in plant movements as an evolutionary behavior. In the last decade of the century a small number of researchers began to probe the cyclical behavior patterns of plant growth, turning their attention to cell division and elongation specifically. The physiologist who synthesized much of this work and who established the key question of the nature of organic rhythms—are they merely adaptations to external stimuli or rather expressions of innate and thus hereditary characteristics?—was the German plant physiologist Wilhelm Pfeffer (1845–1920). Pfeffer had studied with Julius Sachs, the pioneer in establishing plant pathology as an experimental research field, and eventually succeeded him as Germany's leading plant physiologist.[6] Pfeffer's two-volume *Pflanzenphysiologie* was quickly translated and was internationally influential in shaping interest in biological rhythms as characteristic properties of living things, governing behaviors in an era when study of behavior would come to dominate biology and the human sciences—the era of eugenics, Ivan Pavlov, B. F. Skinner, and Aldous Huxley.[7]

Wilhelm Pfeffer's ideas about the nature of biological rhythmicity in plants evolved over the decades that straddled the turn of the twentieth century. His first major exposition of the phenomena in 1875 was groundbreaking, eliciting the comment from Francis Darwin (1848–1925), who had collaborated with his father in the study of plant movements,[8] that "No one has thrown more

---

[6]On Sachs and Pfeffer, see Eugene Cittadino, "Botany," chapter 13 in Peter J. Bowler and John V. Pickstone, eds., *The Modern Biological and Earth Sciences*, v. 6 of *The Cambridge History of Science* (Cambridge: Cambridge University Press, 2009), 225–42.

[7]William Pfeffer's general textbook on plant physiology, *Pflanzenphysiologie. Ein Handbuch des Stoffwechsels und Kraftwechsels in der Pflanze* (Leipzig: Wilhelm Engelmann, 1881), revised as *Pflanzenphysiologie. Ein Handbuch der Lehre vom Stoffwechsels und Kraftwechsels in der Pflanze*, 2 vols. (Leipzig: Wilhelm Engelmann, 1897–1904), was translated into English by Alfred J. Ewart as *The Physiology of Plants. A Treatise upon the Metabolism and Sources of Energy in Plants* (Oxford: Clarendon Press, 1900–1906). These basic textbooks are widely cited by biological rhythms researchers in the first half of the twentieth century, and Pfeffer is regarded as a key author in many scientists' reflections on the development of the field.

[8]Charles Darwin and Francis Darwin, *The Power of Movement in Plants* (London: John Murray, 1880). Bünning, "Opening Address" p. 1, noted that whereas Pfeffer regarded the continued movements of plants in conditions of constant darkness as an after-effect of the previously experienced light-dark alternations, Darwin "clearly emphasized the inherent nature of diurnal periodicity in his book 'On the Power of Movement in Plants' (1880)."

light on the periodic phenomena of plants than Pfeffer, whose admirable 'Periodische Bewegungen' is known to all physiologists."[9] Pfeffer had repeated earlier experiments by De Candolle and was initially reluctant to admit that the causes for rhythmic movements evident in plants might be internal (endogenous), but after reading the work of Joseph Baranetzky and further contemplating his results, by the end of his career he had come to distinguish what he called "aitionomic" or external rhythmicity from "autonomic" or internal.[10]

Pfeffer's analysis of the causes of rhythmicity as either *aitionome* (exogenous) or *autonome* (endogenous) generated commentary in the first decades of the century and set the stage for the theoretical discussion of experimental and observational findings by the next generation of researchers, who focused their attention on biological rhythms.[11] One of the earliest of these was Rose Stoppel (1874–1970), who wrote a doctoral dissertation under Friedrich Oltmanns at Freiburg on the influence of light on the opening and closing of flowers and critically engaged Pfeffer's results and interpretations.[12] Stoppel published this dissertation in 1910 and continued to pursue research and publish on the rhythmicity of plants into the 1940s. Her work was particularly influential for Erwin Bünning, whose long life, persistent research and publication, and participation in international meetings eventually eclipsed Stoppel's

---

[9] Francis Darwin and Dorothea F. M. Pertz, "On the Artificial Production of Rhythm in Plants," *Annals of Botany* O.S. 6 (1892): 245–64, pp. 262–3. The authors also cited Pfeffer's *Pflanzenphysiologie*, and their work should be interpreted in some sense as a reaction to Pfeffer's ideas about plant movement. The Darwins had studied the rotating motion of the ends of plant tendrils, which they believed were responsible for plants' ability to climb. They concluded that this behavior might be the result of an internal rhythmicity that was itself an expression of the inner vitality of living matter: "This repeating power may be that fundamental property of living matter, which stretches from inheritance on one side to memory on the other—a region too wide for the limits of our present paper" (p. 263). Pfeffer had placed rhythmic plant movement in the context of vitalism, but clearly preferred a mechanical, materialist explanation; e.g. Pfeffer, *The Physiology of Plants* (1900), pp. 5–6: "There is no reason for regarding life as the product of an extraordinary and mystical natural force; it is to be treated simply as a special and peculiar manifestation of energy."

[10] Joseph Baranetzky, "Die tägliche Periodicität im Längenwachsthum der Stengel," *Mémoires de l'Académie Impériale des Sciences de Saint-Pétersbourg*, ser. 7, vol. 27 (1879); Wilhelm Pfeffer, *Beiträge zur Kenntnis der Entstehung der Schlafbewegungen* (Leipzig: B.G. Teubner, 1915), cites a reprint. It is Pfeffer's mature ideas that formed the basis for follow-up studies by Rose Stoppel and the Swedes Lars-Gunnar Romell and Martin Stålfelt (see note 12).

[11] One of Pfeffer's contemporaries, who took issue with his conclusions, was Richard Semon, "Über die Erblichkeit der Tagesperiode," *Biologisches Centralblatt* 25 (1905): 241–52 and "Hatt der Rhythmus der Tageszeiten bei Pflanzen erbliche Eindrücke hinterlassen?" *Biologisches Centralblatt* 28(1908): 225–43.

[12] Her doctoral dissertation was published as Rose Stoppel, "Über den Einfluß des Lichtes auf das Öffnen und Schließen einiger Blüten," *Zeitschrift für Botanik* 2 (1910): 369–453. Arthur Jores, "The Origins of Chronobiology: An Historical Outline," *Chronobiologia* 2 (1975): 155–59, p. 155 noted that Rose Stoppel was Pfeffer's student at the University of Strasbourg and that she was "one of the first women admitted to a German university and the first to become a professor of botany in Hamburg," but her connection with Pfeffer, who was not sympathetic to women academicians in his field, was at best casual. See F. Brabec, H. Engel, and H. Söding, "Rose Stoppel 26.12.1874–20.12.1970," *Berichte der Deutschen Botanischen Gesellschaft* 84 (1971): 351–361.

reputation.[13] This might be partly the result of inherent gender inequalities, but it is also the case that follow-up research on the bean *Phaseolus* by Bünning and Kurt Stern in 1928 and 1929 revealed that plants are sensitive to the orange–red spectrum of photographic safe-lights used in darkrooms, which discredited Rose Stoppel's early experimental findings regarding precise diurnal rhythms in plants. Unsuspecting, she had used such lights in the laboratory, and the daily rhythm of switching them on and off for taking measurements was triggering the plant's behavior and vitiating her results.[14]

Besides Stoppel and Bünning a number of other German researchers emerged in the 1930s as leaders in biological rhythms research, chief among them Arthur Jores and Hans Kalmus, both of whom would play a role in the institutionalizing of the research area, influencing several scientists from the low countries and a number of Swedes. German, Dutch, and Belgian research-ers were at the forefront of technological and scientific development during the interwar period, as were to a lesser extent the Scandinavians, whose coun-tries were late to industrialize but benefitted from neutrality during the First World War. Antonia Kleinhoonte was perhaps the most active of the early Dutch researchers, but her countryman Frits Gerritzen (1904–84) also played a role in shaping the early discipline, and the contributions of the Philips

---

[13] For example, biological rhythms researcher Björn Lemmer, "Discovery of Rhythms in Human Biological Functions: A Historical Review," *Chronobiology International* 26 (2009): 1019–68, p. 1022, labels him "one of the founding fathers of chronobiology." Rose Stoppel, "Analyse der tagesrhythmischen Blattbewegungen von Phaseolus multiflorus," *Acta Medica Scandinavica* Suppl. 108 (1940): 45–68. This was her last paper on rhythms before her retirement in 1944. Later scholars seldom refer to her works, citing Bünning's publications instead.

[14] Erwin Bünning, "Opening Address: Biological Clocks," p. 2: "Studies of diurnally periodic leaf movements in the dark room revealed a preponderance of certain movement phases at certain times of day. The cause of this preponderance was, however, eventually discovered. In 1928 and 1929, in the course of a medical research program on the physiological action of atmospheric electricity, my deceased colleague Kurt Stern and I found that these movements were regulated, not by atmospheric electricity or some unknown cosmic factor, but, rather, by the red light in the darkroom [i.e. used by Rose Stoppel in the laboratory]." Bünning published these results in E. Bünning and K. Stern, "Über die tagesperiodischen Bewegungen der Primärblätter von *Phaseolus multiflorus*. II. Die Bewegungen bei Thermokonstanz," *Berichte der Deutschen Botanischen Gesellschaft* 48 (1930): 227–52. It is interesting that Bünning did not cite Stoppel's work in his historical narrative in 1960, but his discovery of her shortcomings was well known. Antonia Kleinhoonte, "Die Tagesperiodik in der Pflanzenwelt," *Deutsche Medizinische Wochenschrift* 64 (1938): 738–42, p. 741, noted "Der Fehlschluß von Rose Stoppel," because of the "Einfluß der roten Beleuchtung für die etiolierten *Phaseolus*pflanzen," without referencing Bünning, but a couple of years later Lars-Gunnar Romell, "Något om rytm och periodism hos växter," *Medicinska Föreningens Tidskrift* 20 (1942): 2–10, p. 7, made the connection clear: "Ungefär samtidigt ansågo sig Bünning & Stern, Rose Stoppels tidigare medarbetare, ha kommit på hemligheten med den tidsinställda dygnsrytmiken hos hennes mörkrumsplantor av bönor. Rytmen befanns ha ett fast fasläge i förhållande till det klockslag, då man gör i ordning bönplantorna för att registrara deras rörelser. Detta är ett tidsödande arbete, som man har brukat göra vid rött ljus i tro att detta icke inverkar . . . (Bünning, 1931)." [Around the same time Bünning and Stern, Rose Stoppel's former collaborators, considered themselves to have discovered the secret of the timed diurnal rhythms of the beans in her darkroom. The rhythm was found to have a fixed-phase position relative to the time when one prepared to record the movements of the bean plants. This is a time-consuming work, which one used to do with a red light, in the belief that it did not affect [the experiment] . . . (Bünning, 1931).]

engineer Balthasar van der Pol (1889–1959) added an important physical-modeling dimension to speculations about the mechanisms of organic timing. That aspect of the field would become especially salient as cybernetic perspectives on biological rhythms came into focus, along with attempts to define the chemical–physical basis of internal timing. Lesser-known names among this generation of researchers are Lars-Gunnar Romell (1891–1981) and Martin Gottfrid Stålfelt (1891–1968), who published their research in German, but their names provide clues to their Scandinavian origin. Both were Swedes, and the fact that they were engaged in biological rhythms research already in the second and third decade of the twentieth century provides context for understanding the role of Swedish leadership in establishing the field as an international, multidisciplinary research endeavor.

3

Biological Rhythms Research in Early Twentieth-Century Sweden

Lars-Gunnar Romell received the doctorate (PhD) at Stockholms Högskola (Stockholm University) in 1922 and was appointed Assistant Professor in plant biology there during 1923–41. Meanwhile, he also received an appointment in the Swedish Department for Forest Research 1918–29 and accepted a newly created position in soils at Cornell University 1928–34. When he returned to Sweden he was appointed to Sweden's College of Forestry (Skogshögskola).[1] Romell's work in biological rhythms was one of the many attempts to follow up on Pfeffer's studies of plant physiology, especially investigating whether observed diurnal rhythms in plant behavior had a basis in endogenous or exogenous metabolic characteristics.[2] Daily variations in the flow of sap in plants had been known at least since the work of Stephen Hales in the early eighteenth century. Romell set out to see whether these were merely responses to the ambient light–dark cycle or if they reflected the *autonome* rhythms that Pfeffer's work had suggested and that were studied by George Karsten (1863–1937), Stoppel, and others in the period around the outbreak of World War I. Romell's findings did not support the conclusion that sap flow follows endogenous rhythms, but his work amply illustrates the currency of rhythmic studies in Swedish higher education, and he came to collaborate in such studies later, as he turned to plant metabolism and forest ecology.[3]

Contemporary with Romell was another Swedish student of biological rhythms, Martin Stålfelt. Stålfelt began study at Stockholms Högskola in 1911 and completed his doctorate in plant physiology in 1921. He was appointed to the botanical faculty in 1922, studied in Holland and Germany in 1922 and 1923, and, with the exception of study at the Boyce Thompson Institute for Plant Research in Yonkers, New York, in 1928, he remained at the Högskola until his retirement in 1959. He was, like Romell, in Sweden at the time of the foundational meeting of the *Societas* in 1937, already an established plant physiologist in Stockholm. Stålfelt's aim was also to follow up on Pfeffer, Karsten, and other German researchers whose work was pointing to autonomous rhythms in plants. His earliest publications indicate that his study at Stockholms Högskola's Botanical Institute was oriented to understanding at the cellular level the nature of biological rhythms in plant growth that had been described in publications by William Kellicott, Pfeffer, Karsten, and

---

[1] "Romell, Lars-Gunnar Torgny," *Svenska Män och Kvinnor: Biografisk Uppslagsbok*, (Stockholm: Albert Bonniers Förlag, 1942–1955), vol. 6, p. 315.

[2] The term "diurnal" refers to the twenty-four-hour terrestrial day, but came also to be used as an antonym to "nocturnal" over time. I use the term in both senses, depending on context for clarity and reflecting common practice. Scandinavian languages distinguish between a *dag* (day) and the twenty-four hour *dygn* (Swedish) or *døgn* (Danish/Norwegian), making the terminology of daily rhythms clearer.

[3] Lars-Gunnar Romell, "Eine neue Anscheinend tagesautonomische Periodizität," *Svensk Botanisk Tidskrift* 12, no. 4 (1918): 446–63.

others.[4] His work concentrated on the study of cell division and the rhythmicity of this phenomenon. He was not the first to pay attention to this (Kellicott, for example, had speculated that cell division exhibited an endogenous rhythm already in 1904),[5] but he focused on it at length, publishing his doctoral treatise in the proceedings of the Swedish Academy of Sciences, the internationally recognized, prestigious organization co-founded by Linnæus.[6] Nevertheless, Stålfelt's contribution must be seen first and foremost as an extension of the German work in the first two decades of the century. Although cognizant of international work by Kellicott and others, he placed his research specifically in the context of the discussion generated by Pfeffer regarding the endogenous or exogenous nature of biological rhythms and the related question about the heritability of behavior, in this case temporal behavior.

Stålfelt's studies of the pea *Pisum sativum* further identified the rhythmicity of cell division in terms of the stages of division and affirmed that this rhythm was endogenous, inasmuch as a diurnal rhythmicity was maintained when the conditions of growth were held constant.[7] This conformed to what Pfeffer had called "autonomic" rhythmicity. Stålfelt's doctoral dissertation indicates that he had concerns about how Pfeffer had defined "autonomic" versus "aitionomic" rhythmicity, specifically what this nomenclature signified in terms of the true causes of rhythmicity and how these were to be parsed out in terms of phylogenic and ontogenic characteristics. He reasoned that all phylogenic traits were imposed by the environment, and that the distinction was therefore most important at the level of ontogeny, in which case one could speak of inherited temporal characteristics versus timings acquired from the environment. The criticism that Pfeffer's work had drawn in the first decades of the century was an indication to Stålfelt that the problems of nomenclature and the experimental

---

[4] Martin G. Stålfelt, "Über die Schwankungen in der Zellteilungsfrequens bei den Wurzeln von *Pisum sativum*," *Svensk Botanisk Tidskrift* 13 (1919): 61–70; Idem, "Ein neuer Fall von tagesperiodischem Rhythmus," *Svensk Botanisk Tidskrift* 14 (1920): 186–89. Both articles are indicated as reported from the Botanical Institute in Stockholm.

[5] William E. Kellicott, "The Daily Periodicity of Cell-Division and of Elongation in the Root of Allium," *Bulletin of the Torrey Botanical Club* 31 (Oct 1904): 529–50, esp. p. 550: "Rhythmic activity resulting from the uniform action of stimuli or forces is quite universal and it may be that here in the root where external conditions are practically uniform, we are dealing with a rhythm which is not related directly to the external environment, but which results from the activity of the root itself, *i.e.*, is internal in its origin."

[6] Tore Frängsmyr, ed., *Science in Sweden. The Royal Swedish Academy of Sciences 1739–1989* (Canton, MA: Science History Publications, 1989).

[7] M. G. Stålfelt, *Studien über die Periodizität der Zellteilung und sich daran anschliessende Erscheinungen*, Kunglige Svenska Vetenskapsakademiens Handlinger vol. 62, no. 1 (Stockholm: Almqvist and Wiksell, 1921), 88: "Wurzeln von *Pisum sativum*, unter 'konstanten Aussenverhältnissen' aufgezogen, zeigen tagesperiodische Schwingungen in ihrer Zellenproduktion mit einem Maximum ungefähr 9–11 Uhr vorm. und einem Minimum 9–11 Uhr nachm. Die Amplitüde der Periode beträgt für Durchschnittswerte einer Population ungefähr +/- 20%." [The roots of *Pisum sativum*, raised under constant external conditions, show diurnal swings in their cell production, with a maximum approximately 9 to 11 a.m. and a minimum 9 to 11 p.m. The amplitude of the period contributes for average values of a population approximately plus or minus 20%.]

evidence itself needed clarification and elaboration, a sentiment that would echo down the century.

In particular, Stålfelt pointed to the need to make a careful distinction between truly endogenous causes of manifest rhythms (i.e., structurally based) and merely *auslösende* or "triggering" causes; namely, exogenous stimuli that provided the occasion for the expression of an innate rhythmicity.[8] A contemporary of Romell and Stålfelt's, Ray Friesner voiced a similar concern for distinguishing the causal factors that governed cell division and elongation.[9] The difference and relationship between fundamental internal rhythms and external triggering causes, which would later be labeled "Zeitgeber" (pacemakers), dominated both the experimental study of and the theoretical explanation for biological rhythmicity as the field developed into the middle of the twentieth century.

But if Stålfelt can also be regarded primarily as an acolyte of Pfeffer's German research program, it is nevertheless important to note that both he and Romell were trained at Stockholms Högskola and that he won appointment to the faculty there, where he remained for many years, helping to transplant Pfeffer's line of inquiry to Swedish academic soil. Stålfelt's research interests eventually drifted away from rhythms studies, and he is best remembered today for his work in plant physiology and ecology.[10] Nevertheless, his work helped give Swedish rhythms research a place in the international scientific community during its formative years. The Germans, at least, knew of his work, as is evident from references to him in Georg Tischler's textbook on plant

---

[8] Ibid., p. 88: "Die Begriffe autonom und aitionom müssen schärfer als bisher definiert werden um die Einwirkung äusserer und innerer Faktoren vollständiger auseinanderhalten zu können, damit in der Diskussion über die Kausalität physiologischer Erscheinungen Missverständnisse vermieden werden, die bei Anwendung der Pfeffer'schen Definitionen unausbleiblich sind. Am einfachsten und am praktischsten dürfte es sein, als autonom nur solche Erscheinjungen oder Vorgänge zu bezeichnen, deren wirkliche (nicht auslösende) Ursachen durch Erbanlagen bestimmt sind." [The concepts autonomic and aitionomic must be defined more sharply than hitherto in order for the effects of external and internal factors to be more thoroughly distinguished from each other, such that in the discussion about causality of physiological phenomena misunderstandings can be avoided, which are unavoidable in the use of the Pfeffer definitions. The simplest and most practical would be to designate as autonomic only such phenomena or processes whose true (not triggering) causes are determined by heredity.]

[9] Ray C. Friesner, "Daily Rhythms of Elongation and Cell Division in Certain Roots," *American Journal of Botany* 7 (1920): 380–407. Unlike Stålfelt and Stoppel, Friesner did not think that electrical potential played a role in the rhythms of cell division and growth, but did point out that the observation that individual research specimens exhibited the same "rhythm" even when individuals might be differently phased implied that the "rhythm" was endogenous and that various external triggering factors were responsible for the "periodicity" (i.e., the phase or location of the underlying frequency in the common time frame).

[10] M. G. Stålfelt, *Växtekologie; balansen mellan växtvärldens produktion och beskatning* (Stockholm: Svenska Bekförlaget, 1960) was translated into English by Margaret S. Jarvis and Paul G. Jarvis as *Stålfelt's Plant Ecology; Plants, the Soil and Man* (London: Longman, 1972). Lars-Gunnar Romell, "Stålfelt," in *Svenska Män och Kvinnor*, vol. 7, p. 309, remembered him especially for developing the Botanical Institute at Stockholm's Högskola (University) into a citadel of experimental research in plant physiology and as a founding member of the Nordisk förening för växtfysiologi (Nordic Association for Plant Physiology).

physiology.[11] That both Romell and Stålfelt carried out elements of Pfeffer's research on rhythmicity in Stockholm in the 1920s and beyond helps explain why the first meeting of the *Societas* was convened in Sweden.

By the middle of the 1930s biological rhythms research of one sort or another occupied a number of researchers of different nationalities, working on diverse subjects, in varied institutional settings, and largely in isolation. At Oxford, for example, Alice Carleton was continuing a line of work on the rhythmicity of cell division pioneered by Kellicott, Karsten, Stålfelt, Friesner, and Joseph M. Thuringer, to name some of the most prominent investigators.[12] Also at Oxford, John Zachary Young had stumbled on the diurnal rhythm of color change in lampreys and had begun systematic experiments to determine the endogenous nature of the timing.[13] By mid-decade the American researcher John Welsh, working at Harvard and at field stations in Cuba and Bermuda, was engaged in a remarkable series of studies of the rhythmic movement of pigment in the eyestalks of crustaceans, which he determined to be controlled by hormones.[14] He acknowledged that Frank A. Brown, Jr., his student, was already doing similar work in 1934.[15] Brown began to work on the daily and tidal rhythms of crabs in the 1940s, mainly at the Marine Biology Laboratory at Woods Hole, Massachussets, but also at Northwestern University in Illinois, and became one of the leading proponents of biological rhythms research in the 1950s and 60s. In Germany, Erwin Bünning, whose name would become almost synonymous with the "biological clock" for his sustained study of the

---

[11] Georg Tischler, *Handbook der Pflanzenanatomie. Allgemeiner Teil: Cytologie* (Die Organe der Zelle), Band II: Allgemeine Pflanzenkaryologie (Berlin: Gebrüder Borntraeger, 1921–22). Under the heading "Die typische Kernteilung" (pp. 252–57), Tischler acknowledged Kellicott's pioneering work in the study of the rhythm of cell division, which Karsten generalized in 1915 and developed in G. Karsten, "Über embryonales Wachstum und seine Tagesperiode," *Zeitschrift für Botanik* 7 (1915): 1–34 and "Über die Tagesperiode der Kern- u. Zellteilung," *Zeitschrift für Botanik* 10 (1918): 1–20. Soon thereafter, Martin Stålfelt began to study the rhythm of nuclear division and pointed to experimental evidence that it was affected by weak electrical currents, an idea that Rose Stoppel followed up in her work on plant rhythms in Germany. Stålfelt's extension of this earlier work was noted by Tischler in the additions he made to his text before publication, pp. 724–25, 730–31.

[12] Alice Carleton, "A Rhythmical Periodicity in the Mitotic Division of Animal Cells," *Journal of Anatomy* 68 (1934): 251–63; Kellicott, "The Daily Periodicity of Cell-Division and of Elongation in the Root of Allium"; Karsten, "Über embryonales Wachstum und seine Tagesperiode"; Friesner, "Daily Rhythms of Elongation and Cell Division, in Certain Roots"; Joseph M. Thuringer, "Studies on Cell Division in the Human Epidermis," *Anatomical Record* 40, no. 1 (Sept. 1928): 1–13.

[13] J. Z. Young, "The Photoreceptors of Lampreys. II. The Functions of the Pineal Complex," *Journal of Experimental Biology* 12 (1935): 254–70.

[14] John H. Welsh, "Further Evidence of a Diurnal Rhythm in the Movement of Pigment Cells in Eyes of Crustaceans," *Biological Bulletin* 68 (1935): 247–52, follows up on his 1930 study using other crustaceans, John H. Welsh, "Diurnal Rhythm of the Distal Pigment Cells in the Eyes of Certain Crustaceans," *Proceedings of the National Academy of Sciences of the United States of America* 16 (1930): 386–95, incorporating Swedish hormone research by Bertil Hanström, Bertil Sjögren, Sven Carlson, and others in Stockholm.

[15] John H. Welsh, "Diurnal Movements of the Eye Pigments of Anchistioides," *Biological Bulletin* 70, no. 2 (Apr. 1936): 217–27, p. 218: "In 1934 Dr. F. A. Brown obtained many more specimens which eventually proved to be *Anchistioides antiguensis* Schmidt."

rhythmicity of peas in particular, continued a line of inquiry begun by Pfeffer and continued at Basel and Hamburg by Rose Stoppel, and in the late 1920s by Antonia Kleinhoonte in Holland.[16] In New York, Frank E. Lutz had determined that crickets' diurnal behavior persists when they are maintained in total darkness.[17] About the same time, Orlando Park and his coworkers were examining the rhythmic behavior of nocturnal insects, following up on similar work with mice pioneered by at the University of Illinois by Maynard S. Johnson.[18] And the German Arthur Jores was working at the University of Rostock with human rhythms.[19] Clearly this was a geographically and topically diverse group, dominated to some extent by Germans. It is therefore somewhat surprising that the meeting convened to form an international society specifically for biological rhythms research was located outside a small town in southern Sweden, a few kilometers inland from the Baltic and remote from Europe's intellectual centers.

---

[16] According to his biographer, Werner Plesse, *Erwin Bünning: Pflanzenphysiologe, Chronobiologe, und Vater der Physiologischen Uhr* (Stuttgart: Wissenschaftliche Verlagsgesellschaft, 1996), Bünning was promoted to the PhD at the University of Berlin in 1929. Erwin Bünning, "Untersuchungen über die autonomen tagesperiodischen Bewegungen der Primärblätter von *Phaseolus multiflorus*," *Jahrbücher für Wissenschaftliche Botanik* 75 (1931): 439–80. Antonia Kleinhoonte, "Über die durch das Licht regulierten autonomen Bewegungen der Canavalia-Blätter," *Archives Néerlandaises des Sciences Exactes et Naturelles*, Série 3B, 5 (1930): 1–110, plus 2 plates.

[17] Frank E. Lutz, "Experiments with Orthoptera Concerning Diurnal Rhythm," *American Museum Novitates* 550 (August 15, 1932): 1–24.

[18] Orlando Park, "Studies in Nocturnal Ecology, III. Recording Apparatus and Further Analysis of Activity Rhythm," *Ecology* 16 (1935): 152–63. Orlando Park and John G. Keller, "Studies in Nocturnal Ecology, II. Preliminary Analysis of Activity Rhythm in Nocturnal Forest Insects," *Ecology* 13(1932): 335–46. Orlando Park, John A. Lockett, and Dwight J. Myers, "Studies in Nocturnal Ecology with Special Reference to Climax Forest," *Ecology* 12 (1934): 709–27. Park and Lockett were at the University of Illinois; Myers at Kent State. Maynard Stickley Johnson's 1926 article "Activity and Distribution of Certain Wild Mice in Relation to Biotic Communities," *Journal of Mammalogy* 7 (1926): 245–77, is widely cited and is remarkable for its early use of clear chronobiological methodology and for putting rhythmicity in the context of an environmental niche.

[19] Jores, "The Origins of Chronobiology," p. 157: "In 1935 I published a summary of the already known circadian rhythms of man, with 274 bibliographic entries." The reference is hindsight; the term *circadian* was not invented until the late 1950s, by Franz Halberg.

4

Why Sweden?

Prompted to record for posterity his recollections of the first meeting of the International Society for the Study of Biological Rhythm, the Bohemian Zoologist Hans Kalmus fondly remembered arriving at Ronneby, Sweden, by train in the evening of 12 August 1937; enjoying the following two days of conversation and conviviality among what he termed "one happy brotherhood" of scientists pursing a common interest.[1] Kalmus, from Prague, had been hiking in Norway with friends, so a train trip to southern Sweden would have been a fairly long, continuous excursion from one remote location in Scandinavia to another. For anybody traveling from elsewhere in the world, the decision to meet near a small town in the largely rural county of Blekinge would have seemed odd, and passing references to the inability of invited American researchers to attend the conference suggest that the location was not chosen for its accessibility to an international group. Why, then, Ronneby? Indeed, why Sweden? What was special about this place at that time?

There are three fundamental factors that help to explain why a meeting to form an international society to study biological rhythms should be convened at Ronneby. First, a number of Swedish academic scientists and physician-researchers were already quite active in various aspects of rhythms research and had formed a small working group to pursue metabolic rhythms in particular in Stockholm, at the Karolinska Institute (Stockholm's medical school and university hospital) and at Stockholms Högskola, which was renamed Stockholm University in 1960. Romell and Stålfelt were at the Högskola, but two physicians at the Karolinska, who were employed at the health spa in Ronneby during the summer of 1937, were no doubt the immediate reason for the meeting's time and place. But this does not explain why Swedes were so prominent in biological rhythms research. To explain this, two additional contextual factors need to be considered. First, there was a history of Scandinavian interest in the effects of light on living things, perhaps a consequence of inhabiting the Far North. Second, there was an intellectual tradition of studying the effects of environmental cycles on organisms that was grounded in the work of Nobel laureate Svante Arrhenius and his colleague Nils Ekholm. I shall treat these in greater detail, beginning with the general geographic and scientific backdrop as a motivation for why there were a number of prominent biological rhythms researchers in Sweden during the 1930s.

The simple fact that Sweden is a land of the Far North, where the annual rhythms of light and dark exert an obvious influence on the physical and psychological condition of people, should be considered a reason for Scandinavians' scientific interest in light and its medical implications. Certainly this is true of late twentieth-century study of seasonal affective disorder and other

---

[1] Hans Kalmus, "The Foundation Meeting of the International Society for Biological Rhythms. Ronneby, Sweden, August 1937," *Chronobiologia* 1 (1974): 118–24.

chronobiological research, but it was also manifest fairly early, beginning perhaps with the work of Anders Ångström (1814–74) on light spectra and the development of light therapy by Niels Finsen (1860–1904).[2] Ångström was one of the pioneers of spectroscopy, closely studying emission lines in the aurora borealis and the solar spectrum, correlating it with the terrestrial emission spectrum.[3] Although he never specifically studied the effects of light on living organisms, his research clearly concerned the interaction of solar radiation with the terrestrial physical environment, implicitly raising the question of the significance of exposure to different spectral components of sunlight.[4] Finsen was trained in medicine at the University of Copenhagen, where he first began to investigate the pathological and subsequently healing effects of portions of the visible and ultraviolet spectrum on patients with smallpox and skin tuberculosis, "inaugurating the modern era of phototherapy."[5] Finsen was put in charge of the newly created Lysinstitut (Institute for Light Therapy) in 1896, which encouraged further research and interest in the healing effects of light. It is my hypothesis that Scandinavia's location at a high latitude, where the annual shift in the length of the day is pronounced and the effect of this on one's sense of well-being is subjectively self-evident, created a sensory and mental context that fostered Swedish interest in correlations between diurnal periods and biological behaviors of all sorts. Ångström won prizes for his scientific work both in Sweden and in England, and Finsen was awarded the Nobel Prize for Medicine and Physiology in 1903, so their work would have been regarded as exemplary by subsequent Scandinavian researchers and should be regarded as relevant social and intellectual context for later scientific work.

Light therapy was still a feature of clinical procedure at St. Göran's Hospital in 1920s Stockholm, and exposure to natural light and fresh air was a commonplace procedure at tuberculosis sanatoria, such as the one in western Sweden headed by Erik Forsgren, the first president of the *Societas*. In some general sense, the importance of light cycles on the daily and yearly rhythms of life was therefore "in the air" in Sweden, and there was an established tradition of Finsen's artificial light therapy within clinical medicine. Such background factors must have supported early interest in the chemical mechanisms underlying physiological responses to light cycles and this makes sense of the occasional

---

[2] Lennart Wetterberg, for example, has had a long career at the Karolinska Institute working on chronobiological problems and was a pioneer in researching the role of melatonin in regulating diurnal rhythms. I thank him for alerting me to the early importance of Ångström and the use of light therapy at St. Göran's Hospital in the 1920s, which got me thinking about Niels Finsen's legacy for clinical phototherapy.

[3] C. L. Maier, "Ångström, Anders Jonas," *Dictionary of Scientific Biography*, ed. Charles C. Gillispie (New York: Scribners and Sons, 1970), vol. I, 166–67.

[4] Anders Ångström, *Recherches sur le Spectre Solaire* (Uppsala: W. Schultz, 1868).

[5] Victor A. Triolo, "Finsen, Niels Ryberg," *Dictionary of Scientific Biography*, vol. IV, 620–21.

references to Niels Finsen one finds in the literature on light rhythms. But beyond the general sensitivity to light and light cycles as important environmental factors, there was a specific research legacy in Sweden that is clearly tied to later biological rhythms research, one that came with the intellectual and academic legitimacy of the Nobel Prize. This was the work of Svante Arrhenius and his lesser-known colleague Nils Ekholm.

Svante Arrhenius (1859–1927) is one of Sweden's most prominent scientists, along with Carl Linnæus, Carl Scheele, and Jacob Berzelius. He is known generally as the first Swedish Nobel laureate for his work on colloidal chemistry, which bridged the turn-of-the-century institutional gap between physics and chemistry.[6] More recently, he has come to historians' attention as the founder of the "greenhouse effect," which is a concept critical to our understanding of global climate change. But Arrhenius's interest in atmospheric gasses and the influence of solar radiation on the biosphere went back to his early enthusiasm for what he called "cosmic physics," borrowing the term from nineteenth-century German authors, but giving it a scope that reached beyond meteorology, to embrace astronomy, physics, chemistry, and the earth sciences in general.[7] The unification of diverse disciplines to study the Earth in a cosmic environment was part of a Romantic enthusiasm for what has been termed Humboldtian science, which characterizes the mentality of cosmic physics. But Von Humboldt's vision of science leaned toward establishing general laws to regulate nature, laws of a more homeostatic character, than did the line of inquiry leading to twentieth-century chronobiology, which probed into the autonomy and variability of such rhythms. Moreover, according to Elisabeth Crawford, Arrhenius developed the concept of cosmic physics in a distinctively Scandinavian context in which the geographic Far North was put to use in developing a science in keeping with the then current ideals of National Romanticism, which interpreted Nordic spaces as special environments. Nevertheless, this late nineteenth-century romantic, Humboldtian science engendered a multidisciplinary approach to the study of nature that is characteristic

---

[6] Elisabeth Crawford, "The Benefits of the Nobel Prizes," in *Science in Sweden: The Royal Swedish Academy of Sciences, 1739–1989*, ed. Tore Frängsmyr (Canton, MA: Science History Publications, 1989), 227–48, p. 237, notes that the first Nobel Institute was created specifically to bridge this gap and provide an institutional home for Arrhenius.

[7] Elisabeth Crawford, *Arrhenius: From Ionic Theory to the Greenhouse Effect* (Canton, MA: Science History Publications, 1996), 132: "In all likelihood, Arrhenius encountered the term 'cosmic physics' in Austria. He first used it in the article in the *Annalen der Physik und Chimie* (1887) in which he reported the results of the experiments he had carried out at Kohlrausch's institute. . . . At this time, Austria was the only place where cosmic physics was institutionalized. The term had originated in Germany, where it was first used by Johannes Müller, . . . in his *Lehrbuch der kosmischen Physik* (1856)." Ibid., p. 133: "In 1888 Arrhenius made good on the promise made in his 1887 *Annalen* article to demonstrate the 'facts of cosmic physics' by extending to atmospheric electricity his laboratory studies of the influence of ultraviolet light on the conductivity of air. At the time this phenomenon was studied most intensively by Franz S. Exner and his circle at Vienna."

of chronobiology and is mentioned on several occasions in the early literature on biological rhythms.[8]

Arrhenius's success in taking cosmic physics from the narrow meteorological definition he had learned in Friedrich Kohlrausch's institute in Vienna and developing it into a multidisciplinary research field was grounded in the intellectual environment at Stockholm's Högskola, which he had helped to foster. The Högskola was established in 1878 and was a fairly new institution when Arrhenius was appointed. On his initiative, a group of scientists from various disciplines created the Stockholm Physics Society, which transformed cosmic physics from a diffuse Humboldtian *Zeitgeist* into a functional research program that entailed the investigation of the cyclical behavior of the cosmos, which provided a unity to the endeavor.[9] It was in this context that Arrhenius began to collaborate with meteorologist Nils Ekholm, who was intent on establishing the terrestrial effects of lunar cycles on a scientific basis, and this in turn led Arrhenius to study biological rhythms.[10] According to Crawford,

---

[8] Ibid., p. 116, regarding the Swedish Academy of Science's expedition to Spitsbergen 1882–83 as part of the First International Polar Year: "In the natural sciences, this focus on the North can be held to have created a 'northern space' that was not merely geographic but represented an important meeting ground for such diverse disciplines as botany, geology, geodesy, hydrography, meteorology, mineralogy, and physics." Such a nexus of disciplines has been widely recognized as necessary for biological rhythms studies. We find mention of the multidisciplinarity of the field already in the opening rhetoric of the first meetings of the Society: Erik Forsgren evoked the multiplicity of disciplines represented at the second meeting (see n. 6, chapter 7, this volume). Arthur Jores, "Eröffnungsaussprache," pp. 16–18 in "Verhandlung der dritten Konferenz der Internationalen Gesellschaft für Biologische Rhythmusforschung," *Acta Medica Scandinavica*, Supplementum 278 (1953), 16, echoed this in opening the third conference (1949): "Die rhythmischen Vorgänge sind in der Natur so weit verbreitet . . . dass es gerechtfertigt ist, wenn Botaniker, Zoologen, Mediziner und Meteorologen sich in dieser Konferenz vereinigt haben, um sich über die Probleme zu unterhalten." [The rhythmic events in nature are so widespread . . . that it is justified if botanists, zoologists, physicians, and meteorologists have come together in this conference in order to talk about the problems.] Arthur Jores, "Zur Rhythmusforschung: Einleitung," *Deutsche Medizinische Wochenschrift* 64, no. 21 (1938): 737–38, p. 738, had made a similar point in his introduction to the papers published in 1938 from the first meeting, but it is not clear that this reflected the opening to that conference.

[9] Crawford, *Arrhenius*, p. 117: "The cosmic physics that came to dominate Arrhenius's science in the 1890s partook of many of the elements that made up the discovery of the nation by its scientists. These elements jelled into a research program (in the broadest sense of the word) rather than remaining part of a diffuse *Zeitgeist* because of the unusual constellation of intellectual partners, institutions, and political beliefs that became his scientific world in Stockholm." Ibid., p. 132: "The aim of cosmic physics was to produce new theories that took into account the interrelatedness of terrestrial, atmospheric, and cosmic events. . . . Cosmic physics at the Society was informed by two ideas. The first, more general one involved the attempt to demonstrate that numerous terrestrial, atmospheric, and cosmic phenomena are cyclical." "Arrhenius was the main promoter of cosmic physics in Stockholm, and work in the field occupied him almost exclusively from the time he left electrochemistry in the mid-1890s until he turned to immunochemistry in the first years of the new century. His work . . . culminated in his *Lehrbuch der kosmischen Physik* (1903). . . . This was very different from his first encounter with cosmic physics in Austria and his first work in the field in the late 1880s."

[10] Ibid., p. 135: "More than any other of Arrhenius's research endeavors, cosmic physics was a group activity resulting from interactions between the members of the Stockholm Physics Society. . . . Hence it was natural that cosmic physics should be strongly interdisciplinary. . . . Out of this intellectual ferment grew Arrhenius's collaborations with Nils Ekholm on lunar cycles."

"Arrhenius's interest in cycles grew out of his studies of atmospheric electricity . . . [his] 'working hypothesis' that the moon was charged with negative electricity."[11] Study of lunar rhythms led him to examine data collected at three of Stockholm's hospitals on the cessation of menses with pregnancy for 10,416 women, from which he drew a correlation between the onset of menstruation and a twenty-six-day cycle in atmospheric electricity that he had discovered.[12]

Arrhenius surmised that electrical tension in the atmosphere, which his work on the northern lights had shown to be a result of solar radiation, was somehow influenced by the lunar cycle and had a physiological effect on humans, which was evident in the menstrual cycle. However, lacking any means of following up on this scientifically, he moved on from biological rhythms research to study glaciation and eventually the influence of carbon dioxide on atmospheric retention of solar energy. His colleague Nils Ekholm, however, continued to study environmental cycles and their effects on the biosphere, focusing especially on the correlation between sunspot activity and terrestrial cycles.[13] Although this might have been fringe science in 1900, it was not as far out there in its time as it would appear from the hindsight of a hundred years later. Alexander L. Chizhevskii (1897–1964), for one, built his entire career in the Soviet Union around the study of sunspot cycles and the attempt to find in corresponding emissions of Z-radiation from deep within the Sun explanations for all sorts of terrestrial behaviors, from cycles of epidemic diseases to major shifts in world religion and political revolutions. His line of research was emulated in the United States by William F. Petersen (1887–1950), a physician at The University of Illinois College of Medicine, in Chicago.[14] Contemporary with the foundational meeting of the *Societas* at Ronneby Brunn, Chizhevskii embarked on a kind of cosmic public health initiative:

> If it is assumed that the Z-radiation lies in the region of radio and ultrashort
> radio waves, the thickness of a metallic screen for the protection of patients

---

[11] Ibid., p. 137.

[12] Ibid., p. 138.

[13] Ibid., p. 139: "For Arrhenius cyclicity was transitory, only indirectly influencing his subsequent work in cosmic physics. He reported the findings of the studies he had conducted with Ekholm in the *Lehrbuch der kosmischen Physik* (1903). . . . For Ekholm, [Otto] Pettersson, and Hugo Hamberg, however, periodicity was a lifelong quest, and they frequently presented their results at meetings of the Physics Society. The meteorologists Ekholm and Hamberg studied the influence of sunspots on weather patterns in the hope of developing a method of long-term weather prediction."

[14] A. L. Chizhevskii, "One Aspect of the Specific Bioactive or Z-Radiation of the Sun," in *The Earth in the Universe*, ed. V. V. Fedynskii (Moscow, 1964), 280–307, trans. Israel Program for Scientific Translations. NASA doc. 1.13/2: F-345 (Jerusalem: Israel Program for Scientific Translations, 1968). Here Chizhevskii lays out his theory of Z-radiation, but also provides some historical background. William F. Petersen, *Man, Weather, Sun* (Springfield, IL: Charles C Thomas, 1947), synthesized a great deal of material on cycles drawn from Chizhevskii and others.

against these radiations is easily calculated. On the basis of these premises I published a series of articles, in the medical press, in 1937, indicating the need to provide all hospitals with electrically grounded, shielded wards impenetrable by Z-rays.[15]

Although scientific reception of Chizhevskii's work was tainted by his insistence on the terrestrial effects of otherwise undetectable Z-rays, he was an important Soviet scientist with an international circle of correspondents, until he was thrown into a gulag by Stalin in 1942. He was the only Soviet member of the International Society for the Study of Biological Rhythm in 1940.[16]

Chizhevkii's interest in Arrhenius's work and in biological rhythms research was not an isolated instance. Both Martin Stålfelt and, following his lead, Rose Stoppel considered the influence of electricity on the rhythmicity of plant physiology. In Stålfelt's case this may have been a direct consequence of the work of Arrhenius and Ekholm at Stockholms Högskola, where Stålfelt was educated. It is therefore not wild speculation to suppose that others who studied in Stockholm in the wake of the Nobel Prize winner would take seriously the idea that biological rhythms manifest in organisms had a physical basis in environmental fluctuations. Teasing out the distinction between seeing these as grounded in organisms as endogenous mechanisms or mere mechanical translations of exogenous stimuli was another matter, however; one that took its lead from the line of inquiry brought by Pfeffer and the Germans. Stålfelt, of course, was party to both streams of thought.

---

[15] Chizhevskii, "One Aspect," p. 299.

[16] "Midgliederverzeichnis," in "Verhandlungen der Zweiten Konferenz der Internationalen Gesellschaft für Biologische Rhythmusforschung am 25. und 26. August 1939 Utrecht (Holland)," ed. Hjalmar Holmgren, *Acta Medica Scandinavica* Suppl. 108 (1940), 11–14.

5

Why Ronneby?

Ronneby in the 1930s was a small town on Sweden's south coast in the Swedish province Blekinge, located on a rail line running east to the important Baltic naval base and shipyard at Karlskrona. What set it apart from similarly sized Swedish towns was the nearby mineral spring, Ronneby Brunn. The term *Brunn*, "well" in Swedish as in German, denotes also wells or springs that were considered to have special healing properties, which the Germans denote by the term "Bad" and the English "bath," as in the case of Bath, but the Belgian name, "Spa," came into the English language for such places more generally. It was discovered in 1705 that the water from a wellspring on the farm Carlstorp, located on the west bank of Ronneby River, downstream from the town, had an unusually high iron content, probably along with other minerals and salts that were thought to be healing when taken internally or used for bathing. The Ronneby Brunn Company was formed to capitalize on this development.

Sweden was relatively prosperous at the time, having expanded imperially across the Baltic and subdued rival Denmark in the seventeenth century.[1] The eighteenth century was in many respects the beginning of modern Sweden. Although not yet industrialized in the sense that it would become in the late nineteenth and early twentieth centuries, the foundations for Enlightenment mercantilism and a Baconian scientific state were laid. A parliamentary system was gaining a measure of power over the traditional council of aristocrats, the government had become more bureaucratic, and the Swedish Academy of Sciences was founded with the aim of carrying out Baconian reforms. Small industrial and commercial development was beginning to exploit Sweden's natural resources. It had new-won political reach as a maritime power that dominated the Baltic Sea and controlled trade into and out of Western Russia and Swedish possessions in northern Germany. Sweden had few spas at a time when Swedes were becoming wealthier, more urbane, more European culturally, and it was a good time to discover and develop a new spa, especially one that was located in the sunny summer landscape of southern Sweden. Even today, Swedes vie with Danes and a few Germans to repopulate the provinces of Blekinge and Skåne every summer, seeking to escape the noxious busyness of urban living.

The Brunn Company developed facilities to supply and entertain a mixed clientele of wealthy spa-goers, who required lodging, recreation, and entertainment, and less well-off locals, who were in many cases transient or lodged in the local hospital. Because the Brunn's "patients" came there primarily for

---

[1] In fact, Blekinge is one of three provinces of southern Sweden that had long been part of Denmark, but were lost to the Swedish Crown in the Treaty of Roskilde (1648) and are only now, since the bridge linking Denmark and Sweden was opened, being re-occupied by Danes in search of summer homes in the greater Copenhagen area.

medical care (many of them diabetics), they required a small medical staff to advise them on matters of regimen and the administration of the spa water, and to attend to their other health needs. By 1937, this staff consisted of a senior physician, a junior physician, and an assistant physician, supported by various staff who bathed the clients and served the restaurant, laundry, and so forth. It was in the Company's interest to hire socially prominent, academic physicians, and it was in the interest of physicians associated with academic institutions to find summer employment, especially in a pleasant, rural location where one could rub elbows with the wealthy. Therefore, it was not unusual for the Senior Physician, at least, to be recruited from Stockholm's prestigious Karolinska Institute, and this was the case in 1936–37, when Dr. Jakob Möllerström was employed as Senior Physician.

Jakob Möllerström (1894–1971) studied medicine at the University of Uppsala and was licensed as a physician in 1924 and awarded an MD in Stockholm in 1930. Appointed as an assistant (Amanuensis) in medical chemistry at Uppsala University in 1917–18, Möllerström moved toward a clinical career in 1923, when he was employed at Hålahult Sanatorium and became an assistant physician at Saint Erik's Hospital (1926, 1928–31) and the Garrison Hospital (1926–28) in Stockholm, also establishing a private practice there in 1931. In the hospital setting he was able to conduct extensive clinical research on metabolism, especially regarding diabetes.[2]

Möllerström's earliest publications show an interest in temporal variations in blood chemistry, beginning with a 1928 study of the fluctuations of peroxidase. He published seven more papers relating to periodicities in blood fractions prior to 1937, focusing on diabetic patients in particular.[3] His 1929 paper on the daily variations of sugar in the blood and urine of diabetic patients at Saint Erik's Hospital shows that he had already begun to suspect that factors other than alimentary ones played a role in the variability of blood-sugar levels.[4]

---

[2] "Möllerstrom, Jakob Valfrid," *Svenska Män och Kvinnor*, v. 5, p. 384; A. Wistrand, ed., *Svenska Läkare i Ord och Bild: Porträttgalleri med Biografiska uppgifter över Nu Levande Svenska Läkare* (Uppsala: Almqvist and Wiksell, 1939); Stina Holmberg, ed., *Svenska Läkare* (Stockholm: P.A. Norstedt & Söners Förlag, 1959).

[3] Arne Sollberger, "In Memoriam Doctor Jakob Möllerström," *Journal of Interdisciplinary Cycle Research* 3 (1972): 1–2, p. 1: "He [Möllerström] made biochemical studies before biochemistry was recognized as a medical discipline in Sweden." The reference is to a French summary of a paper by Jakob Möllerström, "Action de la peroxydase du plasma sanguin," *Comptes Rendus des Séances et Mémoires de la Société de Biologie* 98 (1928): 1361–64. Arne Sollberger, *Biological Rhythm Research* (Amsterdam: Elsevier, 1965), p. 399 lists Möllerström's chief scientific publications.

[4] Jacob Möllerström, "Om dygnsvariationer i blod- och urinsockerkurvan hos diabetiker," *Hygiea* 91 (1929): 379–98, pp. 379–80: "Redan innan jag hade närmare kännedom om Erik Forsgrens undersökningar, för vilka han nyss redogjort, hade jag av vissa anledningar börjat intressera mig för blodsockerförhållandena hos diabetici och därvid gjort en del iakttagelser, som voro anmärkningsvärda och svårförklarliga, om hänsyn togs blott till alimentära faktorer." [Even before I had detailed knowledge of Erik Forsgren's studies, which he has just explained [at the lecture preceding this one], I had for certain reasons begun to interest myself in the glucose conditions of diabetics and thus had made some observations, which were remarkable and inexplicable, if one onely takes into account alimentary factors.]

Möllerström realized that Forsgren's discovery of rhythms in the liver's production of glycogen implied that diabetics should exhibit rhythmic changes in blood sugar and credited his colleague with "beginning a new phase in the study of the physiology and pathology of carbohydrate metabolism."[5] The importance of the realization that the diabetic's blood-sugar level followed a diurnal rhythm and was in part independent of mealtimes was not lost on the Swedish researcher: The use of insulin in treating diabetics was not yet fully worked out, and many physicians continued to depend on strict control of diet. Möllerström immediately grasped the point that knowledge of endogenous carbohydrate metabolic rhythms is necessary both for interpreting the results of drug trials and for making therapeutic decisions for the diabetic. In particular, it was important to understand the rhythm of the body's native production of insulin when determining a supplementary dosage.[6]

I have found no record of Möllerström having held an academic position prior to his appointment at the Wenner-Gren Institute in 1939, but he must have begun working with the up-and-coming young histologist at the Karolinska Institute, Axel Hjalmar Holmgren (1905–1951), during the mid-1930s. Some connection between the two is suggested by the fact that Möllerström brought Holmgren with him to Ronneby Brunn to fill the position of Junior Physician in the summer of 1937. Hjalmar Holmgren, who went by his middle name, had studied in Stockholm, where he was licensed in medicine in 1934 and attained the MD in 1936. He had scaled the academic ladder in the Histology Department of the Karolinska Institute, beginning in 1928, becoming Assistant Professor (Docent) there in 1936 and Associate Professor (Prosector) in Anatomy in 1937. Awarded the Karolinska Institute's Florman Stipend in 1936 and the Alvarenga Prize from the Stockholm Medical Association in 1937, he was something of a golden boy when he joined Möllerström at Ronneby Brunn for the summer season of 1937.[7] Möllerström and Holmgren formed a working partnership at the Karolinska Institute to conduct research on the rhythmicity of metabolic functions, both men influenced by the pioneering work of Erik Forsgren.

Erik A. Forsgren (b. 1896) studied at the Karolinksa Institute (MD 1927) and was appointed to teach histology there in 1927.[8] He was appointed Junior Physician at Hässleby Sanatorium in 1922–23 and Sundsvall Sanatorium in 1924 before his appointment as Lecturer and then Assistant Professor in Histology at the Karolinska in 1927–31. During this time he also taught at Saint

---

[5] Ibid, p. 390: "Forsgrens fundamentala histo-fysiologiska undersökningar över leverns rytmiska funktion kunna därför utan tvekan anses som början till en ny fas i studiet av kolhydratsättningens fysiologi och patologi."

[6] Ibid., pp. 392, 396.

[7] "Holmgren, Axel Hjalmar," in *Svenska Läkare i Ord och Bild*, ed. Wistrand.

[8] "Forsgren, Erik Abraham," *Svenska Män och Kvinnor*, vol. 2, p. 559; Wistrand, *Svenska Läkare i Ord och Bild*.

Erik's Hospital (1927–29) and was appointed Junior Physician there from 1929–31. So, he had first-rate academic training and ample hospital experience when he was appointed Senior Physician at the large sanatorium at Svenshögen in western Sweden in 1931 and moved away from the eastern centers of Swedish academic life in Stockholm and Uppsala. But what he lost in proximity to his colleagues, he gained by having a large patient population to study, mostly suffering from tuberculosis, but also from diabetes and other diseases.

During his studies of the metabolic function of the liver in Stockholm during the late 1920s, Forsgren discovered that the diurnal rhythm evident in the production of bile and glycogen corresponded to observable changes in the liver tissues of rabbits and therefore provided a morphological (histological) basis for rhythmic functions.[9] Forsgren's work, though in many respects a logical development of the line of research on glycogen metabolism that Claude Bernard had begun in the nineteenth century, was widely cited both within Sweden and beyond, and his place as a pioneer in modern biological rhythms research is secure on that basis. Certainly his work provided a stimulus to the research of Möllerström, Holmgren, and others in Stockholm during the 1930s and 1940s, particularly Hjalmar Holmgren's histological studies.[10] Forsgren's position at Svenshögen Sanatorium enabled him to continue the biological rhythms research he had begun at the Karolinska Institute, and he developed and exploited his initial discovery during ensuing decades.[11] His prominence as an early pioneer would sanction his presidency of the *Societas* when it was created at Ronneby Brunn, and although he does not appear to have been as active as Möllerström and Holmgren in the Society's governance, he remained its president for a decade and a half. I have not yet located correspondence among Möllerström, Holmgren, and Forsgren, but the idea of forming an international society and having its first meeting in Sweden must have been generated through personal communication among them and

---

[9] Erik Forsgren, "Mikroskopische Untersuchungen über die Gallenbildung in den Leberzellen," *Zeitschrift für Zellforschung und Mikroskopische Anatomie* 6, (1927–28): 647–88; Erik Forsgren, "Om levers rytmiska funktion och insulinets inverkan på levers glykogenhalt," *Hygiea* 91 (1929): 369–78, p. 373: "Vid undersökningar, som jag företagit med assistens av medicine kandidaterna Wilander, Ågren och Hj. Holmgren, har det befunnits, att levers glykogenhalt hos normala kaniner varierar mellan 1% och 13%." [Through research, which I conducted with the assistance of medical candidates Wilander, Ågren and Hjalmar Holmgren, it was discovered that the glycogen content in the livers of normal rabbits varied between 1% and 13%]. He also acknowledged his collaborators in Erik Forsgren, "Über die Einwirkung des Insulins auf die Leber," *Acta Medica Scandinavica* 70 (1929): 139–49.

[10] For example, Heinz von Mayersbach, "An Overview of the Chronobiology of Cellular Morphology," in *Biological Rhythms and Medicine: Cellular, Metabolic, Physiopathologic, and Pharmacologic Aspects*, eds. Alain Reinberg and Michael H. Smolensky (New York: Springer Verlag, 1983), 65: "The high-amplitude circadian rhythm in livers of animals fed *ad libitum*, first observed by Forsgren in 1928 using chemical determinations and by Holmgren in 1936 using histological stainings, can be regarded as the historical initiation of modern chronobiology."

[11] For a list of Forsgren's publications pertaining to biological rhythms research, see Sollberger, *Biological Rhythm Research*, 355–56.

with the chief researchers in Germany and Holland who also came to Ronneby and were active members of the Society from the outset. But besides these three Swedish chiefs—Möllerström, Holmgren, and Forsgren—there were other Swedes who collaborated with them on rhythm research, undertook their own projects, or worked on the periphery. Karl Gunnar Ågren, Olof Mathias Wilander, and Hjalmar Holmgren collaborated with Forsgren during his initial research, and later Ågren, Wilander, and Johan Erik Jorpes (who discovered heparin) extended Forsgren's research on the metabolism of glycogen in rabbits to rats, showing that the natural rhythms of liver function bore directly on clinical research and therapy regarding insulin.[12]

Ågren, Wilander, and Jorpes were all colleagues of Jakob Möllerström at the Karolinska Institute at the time of the Ronneby Brunn meeting. Gunnar Ågren (b. 1907) became a medical candidate in Stockholm in 1928, receiving the MD in 1936. In 1934 he was appointed to teach physiological chemistry at the Karolinska Institute and developed a research specialty in endocrinology.[13] Olof Wilander (b. 1907) was the same age as Ågren and likewise studied medicine in Stockholm (MD 1939). He taught general chemistry at the Karolinska Institute beginning in 1936, and was appointed to the Department of Pharmacology there in 1937. His research specialty was, like Forsgren's, the metabolism of the liver, and he worked with Erik Jorpes on the preparation of heparin.[14] Erik Jorpes (1894–1973) was somewhat older than Ågren and Wilander. Born in Åland, the archipelago stretching across the Gulf of Bothnia between Sweden and Finland, he began medical study in Helsinki, but transferred to Stockholm after two years (in 1920), earning the MD there in 1928. He taught as an assistant in chemistry at the Karolinska Institute in 1924–28 and was awarded a Rockefeller Fellowship to study at the Rockefeller Institute in New York in 1929, after which he was appointed Associate Professor (*Laborator*) in chemistry and pharmacy at the Karolinska Institute. He worked on a wide variety of biochemical processes during his career, including glycogen metabolism in the liver, the chemistry of blood groups, insulin, heparin,

---

[12]Gunnar Ågren, Olof Wilander, and Erik Jorpes, "Cyclic Changes in the Glycogen Content of the Liver and the Muscles of Rats and Mice. Their Bearing Upon the Sensitivity of the Animals to Insulin, and Their Influence on the Urinary Output of Nitrogen," *Biochemical Journal* 25 (1931): 777–85, p. 778: "In order to check Forsgren's results, we conducted experiments on a rather large scale, analyzing the glycogen content of the livers of a considerable number of mice and rats at different times of day. . . . We thought that the diminished sensitivity of the animals to insulin in the evening might be due to an accumulation of glycogen in the liver. This possibility had been suggested by Forsgren himself [1929]. This periodicity, if established would moreover have an important bearing on our views of carbohydrate metabolism. . . . By our experiments on mice and rats we confirmed Forsgren's findings so far as concerns this periodicity in the accumulation of glycogen in the liver." And p. 779: "In the experiments with insulin the increase in the resistance of the mice to the hormone was found to coincide with the accumulation of glycogen in the liver."

[13]"Ågren, Karl Gunnar," *Svenska Män och Kvinnor* 8, 487–88.

[14]"Wilander, Olof Mathias," *Svenska Män och Kvinnor* 8, 382.

nucleic acids, and the diagnosis of diabetes.[15] Nils Ragnar Nilsson (1903–1981) and Måns Arborelius (b. 1896) were also part of the group of researchers surrounding Jacob Möllerström and were named *revisors* (auditors) of the *Societas* from the beginning. Nilsson began study at Stockholms Högskola after completing his preparatory education in 1922 in Karlstad, not far from Ronneby in southern Sweden. He was granted the PhD and began teaching biochemistry at Stockholms Högskola in 1930, was promoted to *Laborator* in chemistry 1933–36, and then Associate Professor (*Docent*) in microbiology at the agricultural college in 1937.[16] Måns Arborelius began study of medicine in Stockholm in 1918, obtaining the MD there in 1930. From 1920 to 1922 he worked as an assistant in physiology at the Karolinska Institute, before taking a series of clinical positions at various hospitals in Stockholm and Sollefteå Military Hospital (1931–36), before being appointed Chief Physician in Halmstad in 1936.[17]

Bertil Hanström (1891–1969) also should be considered part of the group, on account of his work in endocrinology. He received his doctorate in zoology at Stockholms Högskola in 1920 and was appointed to the zoology faculty at the University of Lund in southern Sweden in 1925. During the 1920s and 1930s he made several zoological research trips abroad, twice to the United States, and is best remembered for his work in invertebrate endocrinology.[18] Interest in endocrinology in Sweden during these decades should be considered a supporting factor for biological rhythms research, especially at the Karolinska Institute, inasmuch as the majority of the early rhythms research involved the endocrine system. Swedes and Danes were at the forefront of endocrinology in the post-World War I period and were quick to obtain and test insulin.[19] It was in this context, and because of long-standing interest in the function

---

[15] "Jorpes, Johan Erik," *Svenska Män och Kvinnor* 4, 116.

[16] "Nilsson, Nils Ragnar," *Svenska Män och Kvinnor* 5, 451.

[17] "Arborelius, Måns Olof," in *Svenska Läkare i Ord och Bild*, ed. Wistrand.

[18] "Hanström, Bertil," *Svenska Män och Kvinnor* 3, 313. In 1939 Bertil Hanström, *Hormones in Invertebrates* (Oxford: Clarendon Press, 1939), summarized the early work on crustacean hormones during the 1930s, which aimed to discern the roles of nerves and fluid transport (humors) in the regulation of pigment change. But clearly, John H. Welsh, "Diurnal Movements of the Eye Pigments of Anchistioides," *Biological Bulletin* 70 (1936): 217–27, p. 225, was familiar with his earlier work, some of which was done at Woods Hole: "The work of Hanström (1931, 1934, 1935) and certain of his students (Sjögren, 1934; Carlson, 1935) has done much to further our knowledge of the sources of the eye-stalk hormone or hormones." According to Sven Carlson, "The Color Changes in *Uca pugilator*," *Proceedings of the National Academy of Sciences of the United States of America* 21 (1935): 549–51, p. 549, he and Hanström conducted research on crustaceans at the Marine Biology Station at Woods Hole during the summer of 1935.

[19] Support for endocrinology in Sweden at this time is also evident in the career of Möllerström's colleague, Axel Westman. Although Westman was not engaged in rhythms research, but rather built his career around researching sex hormones and developing therapies based on them, his work was well funded and helped develop endocrinology institutionally, both at Karolinska and elsewhere in Sweden. See Christer Nordlund, *Hormones of Life: Endocrinology, the Pharmaceutical Industry, and the Dream of a Remedy for Sterility 1930-1970* (Sagamore Beach, MA: Science History Publications, 2011).

of bile and glycogen production, that Forsgren was studying the liver and determined its rhythmic cycle. Moreover, Möllerström had become interested in body rhythms mainly through his study of diabetic patients and was known at the time of the Ronneby Brunn meeting mainly as a specialist in diabetes. Diabetes and the reaction of the body to insulin therapy was largely studied and regulated through urine and blood fractions, and it was the daily variability of these that affected dietary and therapeutic decisions and fostered Möllerström's interest in biological rhythms. Endocrinological research at Karolinska in particular was prominent enough to attract funding from the Rockefeller Institute and eventually the Wenner-Gren Institute, which was established by the Swedish Industrialist Axel Wenner-Gren. The decision to create an institute devoted to cell biology and experimental physiology was made in January 1937, and when the Institute finally opened in May 1939, Jakob Möllerström was appointed to head the department of metabolic research, one of the five main departments. His colleague Ragnar Nilsson headed up the department of physiology, and when he took on another position in 1940, he was succeeded by Gunnar Ågren, another of Möllerström's collaborators. These appointments served both to recognize the importance of Swedish work on the rhythmicity of metabolic functions and to attract funding. Möllerström remained at the Institute until he retired in 1957.[20]

Thus, by the mid-1930s there was a solid core of researchers in Sweden who were working with biological rhythms or in related fields, and the subject was growing. This group provided the intellectual center for founding a new research subdiscipline, the outlines of which were taking shape from the background of widely dispersed individual researchers, the most prominent of whom were located in Northern Europe, England, and the United States. But in the late 1930s, the cost of transatlantic travel and the continued effects of economic depression effectively precluded many American scientists from attending European gatherings. One can imagine that the first meeting in 1937 might have taken place in Germany, where there were several prominent rhythms researchers and also a growing interest in endocrinology, but Germany was already politically volatile, and indeed global tension was on the increase

---

[20] Olov Lindberg, "In Retrospect," in *Wenner-Grens Institut 1937–1959* (Stockholm: Almqvist and Wiksell, 1959), 9–14. The Institute was funded largely by Electrolux founder, Axel Wenner-Gren, but also by a grant from the Rockefeller Foundation.

and the suspicion of a European war was growing as well.[21] The same local newspapers that recorded the Ronneby Brunn meeting were filled with news about the Battle of Shanghai, which began 13 August and lasted late into the fall, and various incidents connected with the Spanish Civil War (1936–39).[22] Germany's support for Franco was well known, and tensions about Hitler's further plans must have been in the back of many people's minds. In such a state of affairs, sunny southern Sweden must have seemed particularly inviting.[23]

---

[21] Franz Halberg and Rudolf Engel, "Arthur Jores, in Appreciation," *Chronobiologia* 1 (1974): 113–17, p. 114, noted that Jores, one of the founding members of the society and clearly in contact with the Swedish core group in Stockholm, would after the war develop his laboratory at Hamburg into one of the leading centers for endocrinological research and establish the German Society for Endocrinology (1953), which he presided over from 1953 to 1963. During this same period he also served as president of the International Society for the Study of Biological Rhythm (1949–67). In retrospect, it is plausible that the Society would have been founded in Rostock or Hamburg and that Jores would have played a more central role in its early development, had it not been for the crisis in Germany.

[22] *Ronneby-Posten: Tidning för Ronneby stad och mellersta Blekinge*, 17 August 1937, p. 4, has a story on "Lufstriderna över Sjanghai fortsätta" [Air Combat above Shanghai Continues] and another speculating on Japan's real goal with its wars, next to an article "Spansk ångare torpederad vid Dardanellerna!" [Spanish Steamer Torpedoed in the Dardanelles!] reporting that a submarine flying Franco's flag had torpedoed and sunk the Spanish ship *Ciudad de Cadix*. Just next to the article on the biological rhythms conference in *Blekinge Läns Tidning*, no. 187 (16 augusti 1937), p. 5, an article reported on the Battle for Brunete in Spain and the sinking of the British tanker *George W. McKnight* in the Mediterranean by unknown warships.

[23] Both Jores and Kalmus were clearly disaffected with the German situation. In a later recollection, Hans Kalmus, "The Foundation Meeting of the International Society for Biological Rhythms. Ronneby, Sweden, August 1937," *Chronobiologia* 1 (1974): 118–24, p. 119, reported that in 1937 he was working at Charles University in Prague, a culturally German university in the Bohemian capital of Czechoslovakia, but he is listed as a Professor in London and the Society's only English member in the membership list for the Society for 1939 ("Verhandlungen der Zweiten Konferenz," ed. Holmgren, p. 11). The day Hitler entered Vienna, Kalmus wrote in his autobiography (see note 17, chapter 5), he decided to leave Prague and find a position elsewhere, a decision made further attractive by his expulsion from his position at Charles University that fall. He was offered a stipend in England as a refugee scientist and appointed to the Zoology Department at the Galton Laboratory at University College, London. His relocation and disaffection with Germany at this point is further suggested by a decisive shift in his publication record from German titles to English titles. Jores was dismissed from his position at the University of Rostock, just across the Baltic from Ronneby, for political reasons in 1936 and was not appointed to the University of Hamburg until after the war. See Halberg and Engel, "Arthur Jores," 114.

6

The Meeting at Ronneby Brunn, 1937

The 1937 meeting of northern European pioneers of biological rhythms research at Ronneby Brunn is recognized by many scientists as a formative moment in the history of chronobiology, yet neither it nor subsequent early meetings have been examined in detail by historians. What little attention has been paid to the early development of chronobiology by scientists, popularizers, and by Cambrosio and Keating has focused on the 1960 Cold Spring Harbor meeting and the debate about the nature and direction of the field in the decades that followed it. Therefore, the details of the Ronneby meeting warrant attention: How was it organized? Who were the principal actors? What scientific work did they present? How did members of the *Societas* follow up on this first organizational effort? In the absence of a conference program or full description of the event, we must consider a diversity of sources to reconstruct this important first meeting. Although Kalmus indicated that a conference program had existed, I have not yet located a copy.[1]

Hans Kalmus recorded his arrival at Ronneby: "On the train I met the neurologist Lundquist and his wife, who also came to the meeting. We were met by the late Hjalmar Holmgren of the Karolinska Institute of Stockholm, who was one of the convenors of the Organizing Committee." He recalled that he was about the last of the participants to arrive and that there were about twenty other participants, none of whom he had previously met. This tells us something about the relative isolation in which the early biological rhythms researchers worked, prior to the establishment of an organization to collect them together periodically and seek collective publication of their proceedings. He mentioned that they were all tired from traveling "and dispersed to the single bungalows in the park for the night."[2]

The railroad reached Ronneby in the 1890s, and at some point a spur was laid along the river, extending to the harbor area, and a station built near the farm buildings of Fridhem, just across the narrow river from the Brunn's well house and adjacent buildings. About one and a half kilometers from town, Ronneby Brunn had a rural air about it, but was still within easy walking distance of the amenities of a small town. The local town park had long served the needs of spa attendees, but it was probably not actually this park that Kalmus meant, but rather the spa grounds and surrounding wooded hills that provided recreational space for attendees and gave the place a luxurious rural atmosphere (Figure 6.1). There were various villas and smaller houses on the

---

[1] Kalmus, "The Foundation Meeting," p. 118: "Unfortunately I cannot here [in Africa] lay hands on the Congress Program and must therefore rely on a few diary entries and my own declining memory." Anders Karlsson, archivist for Ronneby Kommun, informs me that few records survive in Ronneby Helsobrunns Arkiv: Ronneby Brunn from the period after 1929, when the Brunn Company went bankrupt, and that there is no program from the 1937 meeting (personal correspondence, 19 August 2011).

[2] Hans Kalmus, "The Foundation Meeting of the International Society for Biological Rhythms. Ronneby, Sweden, August 1937," *Chronobiologia* 1 (1974): 118–24, p. 118.

Figure 6.1 The cover of the 1936 propectus for Ronneby Brunn depicts the Brunn hotel where the 1937 meeting of the *Societas pro Studio rhythmi Biologici* would be held in August 1937 amid an idyllic forested setting overlooking Ronneby River. The inset is superimposed over a depiction of a peaceful Baltic Seacoast scene downstream, complete with tour boat, seagulls, and birches, as if to suggest a continuity between the river and coastline. A sketch of the well house, emblematic of the Brunn itself, reminds the viewer of the spas's role as a health resort.

Photograph courtesy of Ronneby Kommuns Centralarkiv.

spa's grounds and private villas and dwellings along the roadway into town that could be rented.[3] Because the meeting took place near the end of the spa season, it is likely that the conference attendees were lodged in the villas in the park, which would have been coming vacant as the June–August season drew to a close. There was a magnificent hotel at the spa, but it was built mainly as a dining hall and festival complex and consequently had only twelve rooms for private lodging. The "single bungalows" to which Kalmus referred came with maid service (again suggesting that the guests were lodged on the spa grounds), because he tells us that his failure to negotiate Swedish with the chambermaid resulted in her having all his clothes taken away for cleaning, and he was therefore unable to go to breakfast the next morning until Hjalmar Holmgren arrived with a "search party" and some baggy blue shorts he could use. Kalmus said that he was wearing these shorts to open the conference, presumably after breakfast, which set an informal tone for the rest of the meeting.[4] This was Friday, 13 August 1937.

By his own account, then, Kalmus was responsible for opening the meeting and overseeing its main business, which was the establishment of a society to organize future meetings and pursue publication of the results of biological rhythms research. He had been given the assignment of preparing the organizing statutes for the new Society, which he presented at lunchtime on Saturday.[5] The other principal organizers were the local hosts, Jakob Möllerström and Hjalmar Holmgren, and the Hamburg researcher Arthur Jores, who are mentioned in this capacity in the short article written about the meeting for the town newspaper *Ronneby-Posten* the following Tuesday.[6] The article noted that fifteen of Europe's leading specialists in diabetes had convened to discuss the latest methods of treatment, identifying by name Erik Forsgren, Kalmus

---

[3] An article in the local newspaper summing up the season at the spa, "Säsongen vid Brunnen har i sommar varit ovanligt livlig. Kraftig ökning av gästantalet," *Ronneby-Posten: Tidning för Ronneby stad och mellersta Blekinge*, 28 August 1937, p. 3, noted that the spa controlled a large number of rooms for lodging, but that the season had been so successful, especially during June and July, that some visitors rented rooms in private homes nearby: "Särskilt under juni och juli månader var tillströmningen av gäster så stor, att det stora antal rum, varöver Brunnen disponerar, inte på långt när räckte till, utan man var nödsakad att i ganska stor utsträkning anskaffa rum hos privatpersoner, boende i närheten av Brunnen."

[4] Kalmus, "The Foundation Meeting," p. 119: "In these shorts and a shirt I opened the proceedings, which were filmed, and this funny incident broke the ice of Nordic formality and henceforth we were one happy brotherhood." I have thus far been unable to track further mention of any film or photographs.

[5] Ibid., p. 120: "At lunch I presented an outline of the Statutes for the International Society for Biological Rhythms which I had been commissioned to prepare and which was adopted with slight alterations."

[6] "Framstående läkare och vetenskapsmän på besök i Ronneby. Internationell läkarekonferens å Brunnen. Et 15-tal av Europas specialister på sockersjuka konferera," *Ronneby-Posten: Tidning för Ronneby stad och mellersta Blekinge*, Tuesday, 17 August 1937, p. 5: "På initiative av doktor Möllerström, underläkaren vid Ronneby Brunn, docenten Hjalmar Holmgren och doktor Jores från Hamburg hölls i fredags och lördags den första internationella rhythmuskonferensen i världen å Brunnshotellet i Ronneby." [On the initiative of Doctor Möllerström, Junior Physician at Ronneby Brunn, Docent Hjalmar Holmgren, and Doctor Jores from Hamburg the first international conference on rhythms in the world was held on Friday and Saturday at the spa hotel at Ronneby.]

from Prague, A. Kleinhoonte and Gerritzen from Holland, and that the conference was conducted in German.[7]

The fact that *Ronneby-Posten* identified this conference as a *diabetes conference* tell us something about how the public understood the event and, probably, how the conveners explained it to them. Certainly this would not be the way anybody who diligently attended the presentations would have described the conference, which began with a paper on diurnal rhythms in the plant world and—if the presenters followed the order published the next year—ended with a paper on the rhythm of life and psychotherapy. However, it is true that the majority of the presentations were centered around clinical studies and applications of biological rhythms, and diabetes research was a common background. Most likely, Möllerström himself, as Senior Physician at the spa, explained it this way to the reporter.[8] Such an explanation would not only be understood better by the general public, because diabetes was a common disease that was subject to much press following the discovery and deployment of insulin in the 1920s, but would buttress Möllerström's own professional standing at the spa. The *Ronneby-Posten* article on the conference began by reminding the readership of the importance of insulin and scientific medicine in coming to grips with diabetes and brandishing Möllerström's name as an expert in this area.[9] The importance of this kind of advertising to the reputation of the spa is underscored by an article in the paper two weeks after the conference, announcing the end of the season and summarizing its success, which was of great importance for the local economy:

> As is well known, Doctor Jacob Möllerström, the leading and recognized specialist in the area of diabetes, has functioned as Senior Physician. Doctor Möllerström has always had a fully subscribed list of patients and, practically

---

[7] Ibid.: "Konferensen had samlat ett femtontal av Europas förnämsta specialister på sockersjukdomens område, och bland deltagarna märktes förutom de ovan nämda doktorerna Erik Forssgren, Vadstena, Calmus från Prag, Kleinhoonte och Gerritzen fr. Holland m. fl. Under både konferensdagarna höllos ett flertal förnämliga föredrag och diskussioner om de nya rön, man gjort under de senaste åren och om de nya metoder, man börjat använda för att komma till rätta med sockersjuka. Föredragen höllos på tyska och diskussionsspråket var även tyska."

[8] An article printed in the regional newspaper, *Blekinge Läns Tidning* no. 187 (16 augusti 1937), p. 5, is terser than the one in the town newspaper, but contains similar information, suggesting a single author or that it was derivative: "Specialister på konferens i Ronneby. Ett 15-tal läkare från hela Europa deltog. Stor internationell sockersjuksonferens." [Specialists at a conference in Ronneby. About fifteen physicians from all Europe participated. Large international diabetes conference.]

[9] "Framstående läkare," *Ronneby-Posten*, p. 5: "Nya rön ha nu gjorts, och en av de främsta specialisterna på detta forskningsområde har Ronneby Brunn under ett par år haft förmånen att ha såsom överläkare, nämligen med. doktorn Jakob Möllerström i Stockholm, vilken under de somrar han varit knuten till Brunnen betytt synnerligen mycket för denna, och i sommar har doktor Möllerström faktiskt varit överlupen med patienter." [New findings have now been made, and Ronneby Brunn has been privileged to have one of the foremost specialists in this field of research as the Senior Physician for a couple of years, namely, Jakob Möllerström, MD, of Stockholm, who has been especially significant for the spa during the summers he has been linked with it, and this summer doctor Möllerström has actually been overburdened with patients.]

speaking, he has been overwhelmed with work the entire season. Patients have each and all had ample praise for his distinguished medical practice and for his pleasant and amiable manner. Associate Professor Hjalmar Holmgren served as Junior Physician, but the spa also has employed an assistant physician. Associate Professor Holmgren has also had a very extensive practice, and has made himself very popular.[10]

Ronneby Brunn was a private concern that had gone bankrupt in the wake of the 1929 economic crisis and was now operating under new management. Clearly, the professional stature and popularity of its medical staff was important to how Director Aron Johannson, who provided the information for the newspaper article, wished the spa to be seen by the public. The attendance numbers were good, the staff were among Sweden's best trained and most competent, and the continued recovery of the local health business was promising. Unforeseen at that point was the fact that the disruptions caused by the outbreak of World War II, although they did not implicate Sweden as a belligerent nation, would end the spa's significance as a healing venue. Ronneby Brunn thrives today as a sports complex, massage-therapy getaway, and water park. A cup full of the Brunn water is still available to the curious visitor, free of charge, but it comes without medical recommendation.

According to the local newspaper, the conference itself was held in the spa's imposing hotel, an ornate wooden structure designed by a renowned Swedish architect from Malmö and reckoned as the largest wooden building in northern Europe when it was completed in 1897 (Figure 6.2).[11] A conference in such a luxurious, beautiful location must have been restful and conducive to intimate academic discussion, while also showing off Sweden's elite culture to visitors from abroad. Two founding members of the Society, Arthur Jores (Hamburg) and Hans Kalmus (Prague), ethnic Germans, recalled the meeting

---

[10] "Säsongen vid Brunnen har i sommar varit ovanligt livlig. Kraftig ökning av gästantalet," *Ronneby-Posten* (28 August 1937), p. 3: "Såsom överläkare har som bekant den framstående och kände specialisten på sockersjukdomarnas område, doktor Jakob Möllerström från Stockholm, fungerat. Doktor Möllerström har ständigt haft mottagningslistan fulltecknad och praktiskt taget hela säsongen har han varit överhopad med arbete. Patienterna ha samt och synnerligen haft de amplaste lovord för hans förnämliga läkekonst samt för hans trevliga och älskvärda sätt. Såsom biträdande läkare har docent Hjalmar Holmgren tjänstgjort, och dessutom har Brunnen förfogat över en assistentläkare. Docent Holmgren har även haft en mycket omfattande praktik samt har gjort sig mycket populär."

[11] "Framstående läkare," *Ronneby-Posten*, p. 5: "En konferense, som kan få sin historiska betydelse, hölls under fredagen och lördagen å Brunnshotellet i Ronneby." On the history of the hotel, see Elisabeth Mansén, "Brunnsliv & Kurortskultur i Ronneby 1705–2005," pp. 17–82 in *Ronneby Brunn under Trehundra År 1705-2005*, ed. Lis Hogdal (Stockholm: Byggförlaget, 2005), 26, 63.

Figure 6.2 The hotel at Ronneby Brunn, as depicted in a brochure. Photograph courtesy of Ronneby Kommuns Centralarkiv.

at Ronneby Brunn for articles solicited by Franz Halberg for the new journal *Chronobiologia* when it was being established in the early 1970s.[12]

Jores's account is very brief and sketchy on details, but informs us that the initial discussion about organizing a conference was carried out through correspondence and a personal meeting with Möllerström, suggesting the crucial role the two played in the formation of the Society, but Holmgren must also have been involved at an early stage.[13] From 1933 until his dismissal in 1936 on political grounds, Jores was a member of the Department of Internal Medicine at the University of Rostock, which was one of the main Baltic Sea ports connecting Sweden to the continent. There were close commercial and academic ties between Rostock and Scandinavia going back into the Middle Ages, so occasional communication between the two physicians who had a common interest in endocrines and rhythms seems plausible. Jores claimed that seven researchers participated in the conference:

> Forsgren, from Stockholm, the discoverer of the hepatic rhythm; Gerritzen, a physician practicing near Utrecht in Holland, who carried out experiments on the periodicity of water excretion by the kidneys . . . ; Holmgren, university-assistant at the Institute of Anatomy in Stockholm, who had confirmed and extended the reports of Forsgren on the rat and mouse; Kalmus, a biologist from London, who had carried out experiments on circadian periodic phenomena in ants; Kleinhoonte, a botanist, who studied the sleep movements of *Canavalis ansiformis* under usual conditions and in total darkness; and finally, Möllerström, a specialist in diabetes, who tried to apply the concepts of hepatic rhythm to the treatment of diabetics.[14]

This list corresponds well with Hans Kalmus's account in terms of names, but not in terms of detail. Kalmus noted that his account of the meeting was based on his diary notes from the time and on his memory, as he did not have a conference program with him in West Africa, where he was when he wrote his reflection. Nevertheless, his reconstruction of events might be more accurate than Jores's, which is at least marred by his anachronistic identification of Kalmus as an ant researcher from London rather than remembering his early work on bees and *Drosophila*, which was done on the continent. Kalmus had in fact obtained a position at the Galton Laboratory at University College, London, but this was a year after the Ronneby Brunn meeting, at which

---

[12] The first issue came out in 1974, but clearly Kalmus's contribution was solicited no later than 1973, when it was received by the journal. On the foundation of *Chronobiologia* and its role in giving biological rhythm studies a disciplinary identity, see Cambrosio and Keating, "The Disciplinary Stake: The Case of Chronobiology."

[13] See Kalmus's recollection that he was invited by Holmgren, note 16 in this chapter.

[14] Jores, "The Origins of Chronobiology," p. 158.

time he was still in the Zoology Department of Charles University in his native Prague.[15]

Kalmus remembered being invited to help form the new Society by Hjalmar Holmgren,[16] who was no doubt working with Möllerström, and recollected that he opened the meeting Friday morning and that Möllerström subsequently launched the morning session with a welcoming speech, which was followed by papers by Jores, Kleinhoonte, Kalmus, and Gerritzen. He noted that his paper reported on research he had published already in 1935 on the time-sense of bees and the rhythm evident in *Drosophila* eclosion, but his emphasis on this particular work in the early 1970s, when he was asked to record his reminiscences, might also reflect his desire to establish his priority in a line of research that was developed by Colin Pittendrigh in the 1950s and subsequently hailed as a core area of chronobiological research.[17] Kalmus remembered that between the end of the morning session and lunch, conference attendees were invited into a sauna and scrubbed down by spa attendants—a singular experience for the non-Swedes, he noted—and this must have taken another

---

[15] Hans Kalmus and David Benedictus, *Odyssey of a Scientist: An Autobiography* (London: Weidenfeld and Nicolson, 1991), recounted his decision to leave Prague after Hitler's army marched into Vienna, his dismissal from the university, and walking across the border to Hungary in December 1938, with the hope of emigrating to London: "So far as I can ascertain none of my class-mates was powerful in any Nazi organisation, but there were several other of my friends who were. My fellow lecturer in the Zoology Department, with whom I had collaborated on research and climbed in the Alps, was already an SS officer on the day he 'helped' me leave my room after I had been expelled from the university" (p. 15). Kalmus received a stipend earmarked for refugee scientists and was interviewed by G. F. Wells (H. G. Wells's son) at the Zoology Department of University College, London.

[16] Hans Kalmus, "The History and Philosophy of Chronobiology," *Journal of Interdisciplinary Cycle Research* 19 (1988): 227–34, p. 229.

[17] Kalmus, "The Foundation Meeting" (received by the journal in December 1973), p. 119. Pittendrigh came to dominate the public face of rhythms research in the United States from the mid-1950s until his death in 1996, in large part on the basis of his research on the seemingly endogenous nature of the rhythm of *Drosophila* eclosion, and Kalmus may have used Franz Halberg's invitation to reminisce for readers of *Chronobiologia* as an occasion to assert his priority in working with *Drosophila* rhythms and thereby remind his colleagues that Pittendrigh was not the only person in the room. Kalmus did indeed publish results of his research on *Drosophila melanogaster*, in Hans Kalmus, "Periodizität und Autochronie (=Ideochronie) als zeitregelnde Eigenschaften der Organismen," *Biologia Generalis* 11 (1935): 93–114. He used as examples of periodicity the cell division of *Paramecia*, *Drosophila* eclosion (Schlupftermin), and the time-sense of bees. It is clear from his comments that he undertook experiments with *Drosophila* (99): "Die eigenen Versuche wurden mit Drosophila ausgeführt, die sich dafür sehr geeignet erwies." [Our own experiments were performed with Drosophila, which proved particularly suitable.] That Kalmus was recognized for this work at the time is evident from an article contemporary with the Ronneby Brunn meeting: Axel M. Hemmingsen and Niels B. Krarup, "Rhythmic Diurnal Variations in the Oestrous Phenomena of the Rat and Their Susceptibility to Light and Dark," *Det Kgl. Danske Videnskabernes Selskab, Biologiske Meddelelser* 13, no. 7 (1937), 42–43: "Bünning (1935a, p. 608) has adduced evidence to show that the diurnal periodicity in emergence of *Drosophila* imagines from the puparia, as observed also by Bliss (1926), Bremer (1926), and Kalmus (1935), and known also from other insects, is an inherent 24 hours rhythm, which is still retained after keeping this species through several generations in a 16 (8 dark–8 light) hours or 36 (18 dark–18 light) hours rhythm, or in constant faint light."

hour out of the morning, at least.[18] Following lunch there were papers by Forsgren, Holmgren, a second by Jores, and Möllerström. After this, Kalmus wrote, "There was a good deal of formal discussion of the papers which was continued in a lighter vein during dinner, greatly assisted by wine and *aquavit*."[19] This aspect of formal Scandinavian academic meetings has changed little over the years. Presentations continued Saturday morning with papers that Kalmus recalled as "review papers of a more theoretical nature" by Kleinhoonte (plants), himself (animals), and Jores (humans). He noted that these were not published, although the material in his own presentation was similar to a paper he published the following year in the Italian journal *Rivista di Biologia*.[20]

Lacking a conference program, we do not at this time know definitively which papers were presented Friday and which on Saturday morning, but it is possible to reconstruct the likely program on the basis of the recollections by Kalmus and Jores and from the eight papers that were published the following year, without any mention of the conference, in the 20 May and 8 July issues of the weekly professional journal published in Leipzig, *Deutsche Medizinische Wochenschrift*.[21] The records of the *Societas* that were published with the proceedings of its second meeting identified these papers as, with one exception, the proceedings of the 1937 conference.[22] Apart from the two written by Måns Arborelius and J. H. Schultz, the articles were written by the authors identified by Jores and Kalmus as conference participants, but the

---

[18] Kalmus, "The Foundation Meeting," p. 119: "Before lunch we all had a Sauna bath and were scrubbed, completely naked, by elderly female bath attendants—a unique experience for everybody except the Swedes." If this sauna was done collectively, one wonders what Kalmus's Dutch colleague, Antonia Kleinhoonte, made of the experience. I have no information regarding a sauna in the spa complex, but it may have been built into Bathhouse #1, which flanked the spa pavilion, where spa visitors gathered to drink the water. Of course, the water was what it was all about, and the patients could drink the famed local spa water, which was very rich in iron sulphate, or blend it with water imported from other mineral springs. Baths and massages were also a part of the regimen, and Elisabeth Mansén, "Brunnsliv & Kurortskultur i Ronneby 1705–2005," pp. 21–24 recounts the possibilities: "Vid Ronneby brun kunde man bli serverad en rik uppsättning välkända mineralvatten och avnjuta alla upptänkliga badtyper. De gamla gyttjemassagebaden fanns kvar sedan 1700-tallet, likesom stålbaden i järnhaltigt vatten där det finaste kallades Nauheimerstålbad." [At Ronneby Brunn one could be served a rich assortment of well-known mineral waters and make use of all conceivable types of bath. The old mud-massage bath persisted since the 1700s, and also steel baths with iron-rich water [could be had], the finest of which was called the Nauheim steel bath.] Arthur Jores, "The Origins of Chronobiology," p. 158 noted that the Utrecht meeting began with a communal bath in a lake.

[19] Kalmus, "The Foundation Meeting," p. 119.

[20] Kalmus, ibid., noted that this work appeared in Hans Kalmus, "Ueber das Problem der sogenannten exogenen und endogenen, sowie der erblichen Rhythmik und ueber organische Periodizitaet ueberhaupt," *Rivista di Biologia* 24 (1938): 191–225. As the title indicates, he was concerned with problems that were theoretical in nature, especially the central question of endogenous versus exogenous causes.

[21] *Deutsche Medizinische Wochenschrift* 64 (1938): 737–48, 989–98.

[22] The report of the Society's council, published with the proceedings of the second meeting, "Vorstandsbericht für die Jahre 1937–39," pp. 15–18 in Holmgren, ed., "Verhandlungen der Zweiten Konferenz," p. 15, recorded that the proceedings of the first conference were with one exception printed in *Deutsche Medizinische Wochenschrift* nos. 21 and 28 (1938).

order of publication does not correspond to what Kalmus remembered.[23] There is no paper by Kalmus, but because he had already published on the subject in 1935, he may not have prepared one reflecting his Friday-morning presentation. Even if he had, it would not have been printed with the others, owing to a prohibition against publishing the work of non-Aryan authors that was then in effect in Germany.[24] If the report of one paper being published in the *Deutsche Medizinische Wochenschrift* that had not been presented at the conference is accurate, this must refer to either Arborelius or Schultz. It is plausible that Måns Arborelius attended the meeting, because he was a member of the Swedish contingent that was working on biological rhythms. But, neither Jores nor Kalmus recalled his presentation, and I have encountered no evidence to support his presence at Ronneby Brunn. Likewise, there is no evidence of Schultz, from Berlin, having attended the meeting, and his article might have been added as a practical solution to accommodating a paper on a topic that fit neatly with the others from Ronneby.

Arthur Jores of Hamburg wrote an introduction to this first set of papers, suggesting that he mediated the publication, but as the papers were paid for by the authors, they cannot therefore be seen as the Society's publication. Jores began his introduction with a quotation from Ludwig Klages (*"Der Rhythmus aber ist die Urerscheinung des Lebens"*), putting the question of rhythms vs. cycles into a vitalist vs. mechanist context, rhythms being characteristic of living organisms, whereas cycles characterize the machine and the mechanization of modern living.[25] After this romantic and philosophical

---

[23]The papers published under the heading "Zur Rhythmusforschung" in *Deutsche Medizinische Wochenschrift* 64, no 21 (1938; 20 May): 737–48 are, in order, A. Jores, "Einleitung"; A. Kleinhoonte, "Die Tagesperiodik in der Pflanzenwelt"; Erik Forsgren, "Die Rhythmik der Leberfunktion und des Stoffwechsels"; H. Holmgren, "Leberrhythmus und Fettresorption"; F. Gerritzen, "Der 24-Stundenrhythmus in der Diurese"; and in no. 28 (8 July): 989–98, A. Jores, "Endokrines und vegetatives System in ihrer Bedeutung für die Tagesperiodik"; Jakob Möllerström, "Die therapeutische Bedeutung der menschlichen Rhythmik"; Måns Arborelius, "Die klinische Bedeutung der menschlichen Rhythmik"; A. Jores, "Die Ursache der Rhythmik vom Gesichtspunkt des Menschen"; and J. H. Schultz, "Lebensrhythmus und Psychotherapie."

[24]Kalmus commented on his difficulty publishing scholarly papers as the Nazi authorities tightened their grip on German academic expression in Kalmus and Benedictus, *Odyssey of a Scientist*, p. 138: "Before leaving Prague I had refused to publish in the *gleichgeschalteten*, conforming German journals, which would probably have refused to publish me in any case, and as a result I had to turn to obscure Italian and Austrian periodicals." Although raised as a Lutheran and self-identifying as a Christian, Kalmus was of known Jewish extraction and would likely have been killed in the Holocaust, as were his parents, had he not fled Prague for London in 1938.

[25]Arthur Jores, "Zur Rhythmusforschung: Einleitung," *Deutsche Medizinische Wochenschrift* 64, no. 21(1938): 737–38, p. 737: "Takt finden wir daher nur bei vielem, was der menschliche Geist geschaffen hat. Takt zeigen die Maschinen und Motoren, es takt die Uhr an der Wand, und auch das Leben des Arbeiters an der Maschine wie das Leben des Großstadtmenschen ist heute in das Gleichmaß eines eintönigen Taktes eingezwängt. 'Der Rhythmus aber ist die Urerscheinung des Lebens' (Klages)." [Therefore we find 'cycle' only in many of the things that the human mind has created. Machines and engines have cycles, cycle is the clock on the wall, and also the worker's life at the machine, as the life of city dwellers is now wedged into one monotonous cycle. 'Rhythm, however, is the essential manifestation of life' (Klages).]

introduction, Jores provided a bit of historical background to motivate the importance of the subject.[26] He defined two approaches to the study of biological rhythms: The first is the phenomenological; the phenomena themselves must be identified and studied. The second is the physiological; research must be undertaken to determine the significance of biological rhythms for the inner constitution of the body. Perhaps Jores was conscious that his audience, professional physicians and clinical researchers, would be more interested in the practical significance of biological rhythms research for medicine and less in studies of bean leaf movements of the sort that Rose Stoppel was doing and that Antonia Kleinhoonte reported extensively in the paper immediately following Jores's introduction. Perhaps for this reason, too, he did not mention the formation of a new society devoted to biological rhythms research or report on the meeting at Ronneby. But another possibility is that this introduction was what Jores presented at the conference itself, in which case the Society had not yet been formed. But if so, one must construe this in light of Hans Kalmus's recollection that *he* opened the conference proceedings at Ronneby, and that Möllerström had introduced the first session.[27]

Following Jores's introduction, there were four articles published in issue 21 of *Deutsche Medizinische Wochenschrift* and another five in issue 28, two of which were by Jores, corresponding to the two papers Kalmus attributed to him on Friday. The remaining papers are consistent with Kalmus's recollection that Saturday's more theoretical papers were not published, except that his was substantially similar to one he subsequently published in *Rivista di Biologia*. Because the authors bore the costs of publication, this may also have affected which papers were included. In this regard it is interesting to note that only the first paper, by Antonia Kleinhoonte, included a separate references section. It is also by far the longest of the papers published, and this might reflect better financial support for publication or better opportunity to revise the paper before publication, or perhaps she was just more meticulous in her scholarship.

The presenters of record who can be reconstructed in this way cannot have accounted for all of the participants at the meeting, however. Both the *Blekinge Läns Tidning* and *Ronneby-Posten* had recorded about fifteen, and Hans Kalmus remembers meeting about twenty persons other than the two he had met on the train, a number agreeing with the Council report given at the second meeting, which, along with the proceedings of that meeting, constituted the

---

[26] Ibid: "Doch ist erst heute die Zeit dafür reif geworden, den Rhythmus als solchen zum Gegenstand der Forschung zu machen. . . . und so zweifle ich nicht daran, daß die Erkenntnis der fundamentalen Bedeutung der Rhythmusforschung für viele Fragen des Lebens sich bald Bahn brechen wird." [But only now has the time become ripe to make rhythm as such an object of research. . . . and so I have no doubt that the knowledge that rhythm research is of fundamental importance to many of life's questions will soon lead the way forward.]

[27] Hans Kalmus, "The Foundation Meeting," p. 119.

first published account of the new organization. It is possible that the number reported by the newspapers was provided by one of the conference hosts and therefore included *visitors* and did not take into account those already at the Brunn, namely, Möllerström, Holmgren, and others closely associated with them. This group of twenty or so attendees at Ronneby constituted the new Society and appointed a governing council consisting of Forsgren, Möllerström, Jores, and Holmgren, with Arborelius and Nilsson appointed as auditors. Gerritzen was charged with arranging for a second meeting in 1939, in Holland, place to be determined.[28] In addition to these, two alternates ("suppleants") to the council were named, Kleinhoonte and Kalmus.[29] The founding members must therefore have included Jores, Kleinhoonte, Forsgren, Holmgren, Gerritzen, Möllerström, Arborelius, perhaps Schulz from Berlin, and of course Hans Kalmus.

Given the arguments about the direction and funding for chronobiology in the second half of the twentieth century, which sometimes divided along the lines of pure biological science versus clinically driven medical science, it is interesting to note that the large majority of the papers presented at this first international congress pertained to human physiology or were based on clinical studies. This, of course, reflects in large measure how such research was being funded in Sweden, where opportunities for medical scientific research were largely funded through hospitals and associated institute positions (in the case of Karolinska and Wenner-Gren), but it is a point of fact that Erik Forsgren and his medical colleagues were in attendance, and not Martin Stålfelt or Erwin Bünning or others doing cell work, studies on crustaceans and insects, or even plants, with the notable exception of Antonia Kleinhoonte. Because we lack correspondence for the major organizing persons, we cannot know who else might have been invited but failed to come. Both Kalmus and the provincial newspaper recorded that a number of Americans had been invited, but did not arrive, whether because of funding or timing.[30] We can speculate about who they were on the basis of those who had joined the

---

[28] "Vorstandsbericht für die Jahre 1937–39," in "Verhandlungen der Zweiten Konferenz," ed. Holmgren, p. 15. This official but retrospective account agrees with what was reported at the time of the Ronneby Brunn conference in "Framstående läkare," *Ronneby-Posten*, 17 August 1937, p. 5: "I samband med konferensen bildades en internationell rhythmusförening, vilken kommer att sammanträda om 2 år i Holland." [In connection with the conference an international rhythm association was created, which will convene in two years in Holland.]

[29] "In Ronneby 1937 gewählter Vorstand der Gesellschaft," in "Verhandlungen der Zweiten Konferenz," ed. Holmgren, p. 9.

[30] The newspaper article, "Specialister på konferens i Ronneby," *Blekinge Läns Tidning*, 187 (16 August 1937), p. 5, noted that the Americans did not arrive in time for the conference ("Tyvärr hade dock ett antal representanter från U.S.A. ej hunnit fram i tid till konferensen."), leaving open the possibility that they came to Europe, but did not reach the conference. However, Kalmus, "The Foundation Meeting," p. 119, said that some Americans were invited, but could not obtain funding.

Society by 1940, but we cannot know for sure.[31] Whether those papers presented at Ronneby Brunn were representative of the general work being done in rhythm studies or not, it is clear that both Hans Kalmus and the local newspaper believed that the research was significant. Looking back about 25 years later, Kalmus wrote that "In these papers many of the fundamental questions concerning the nature of biological rhythms were posed, some of which are still unresolved and exercise the minds of many research workers in our field." *Ronneby-Posten* described the meeting as possibly of historical significance.[32]

The meeting formally concluded Saturday night with a lavish Swedish-style banquet that Hans Kalmus fondly recalled many years later:

> The afternoon was free for informal discussion, and at 8 p.m. we all assembled for a ceremonial crayfish supper in the Swedish tradition, myself in a newly ironed suit. There is an entry in my diary mentioning the convivial atmosphere created by the steaming food and the many differently coloured drinks. There were numerous speeches including a proposal of mine to introduce the crayfish as a mystical heraldic symbol for researchers on biological rhythms—I was too drunk to realize the irony of this choice, as everybody knows that the crayfish frequently moves backwards.[33]

Apparently there was no structured part of the meeting on Sunday, perhaps breakfast for those rising early enough, and Kalmus recorded only that "The next day the Congress dispersed and contact between members was only resumed before the second [sic] Meeting some time after the war."[34] Perhaps it was a combination of faulty recollection and his own absence from the 1939

---

[31] "Mitgliederverzeichnis" pp. 11–14 in "Verhandlungen der Zweiten Konferenz," ed. Holmgren, p. 14: Prof. Th. H. Bissonnette, Hartford; Asst. Prof. of Anatomy A. L. Grafflin, Boston; S. S. Lichtman, MD, New York; Prof. W. Petersen, Chicago; and J. H. Welsh, PhD, Cambridge, Mass. Kalmus, "The History and Philosophy of Chronobiology," p. 230, recalled that "a couple of colleagues from Haward [sic] could not come because there was no money available!" and, by inference from the above list, those were probably Welsh and Bissonnette.

[32] Kalmus, "The Foundation Meeting," p. 119; "Framstående läkare," *Ronneby-Posten*, 17 August 1937, p. 5: "En konferense, som kan få sin historiska betydelse" [A conference that can have historical significance], and "De i konferensen deltagande läkarna ansågo konferensen synnerligen betydelsefull." [Those physicians taking part in the conference regard the conference as clearly significant.]

[33] Kalmus, "The Foundation Meeting," p. 120.

[34] Ibid. I have no evidence to suggest that Kalmus attended the 1939 meeting at Utrecht, by which time he was living and working in London, and likely confused *his* second meeting, in Hamburg 1949, with the second meeting of the Society. It is somewhat odd that this was not corrected by the editor, but then the purpose of the reminiscence was connected with legitimizing the new journal, *Chronobiologia*, and not formally historical, and Kalmus's purpose, judging by the rest of the article, was to situate his own work within the development of key ideas in the field, namely, the early use of *Drosophila* as a model organism, the imposition of the language of sine-wave oscillations, and the implicit linking of mechanical models to biological phenomena, the suggestion that parts of plants are likely to exhibit rhythmicity, whereas complex animals are more likely governed by "a central *clock* of a neuro-humoral nature" (p. 123), and the claim that the endogenous or exogenous nature of the clock was itself poorly defined and conceived. All of these were more clearly important from the hindsight of 1974 than they were evident from the perspective of 1937.

meeting, but clearly it had been decided at Ronneby Brunn that the second conference should be held in Holland, as the local newspaper mentioned this just days after the conference ended.

7

Swedish Neutrality and Leadership of the International Society for the Study of Biological Rhythm in the 1940s and 1950s

The second meeting of the International Society for the Study of Biological Rhythm was convened in Utrecht, 25–26 August 1939. If there had been hints that all was not well in Europe at the time of the 1937 meeting, the atmosphere must have been positively tense for the second one. Two days before the meeting began, Hitler and Stalin agreed to partition Poland (23 August), and Germany launched its offensive by the end of the next week (1 September), triggering World War II. The Society's planning had gone on in spite of the political tensions in Germany and the hostilities in the Far East, and, with the exception of Arthur Jores (from nearby Hamburg) and Hans Kalmus (German, but now a resident of London), the officers elected in Ronneby were from relatively neutral nations. Of the nine named in the published records of the *Societas* from the Utrecht meeting, the majority (five) were from Sweden, including the president, vice president, the secretary-treasurer, and the two auditors; and two were from Holland.[1] This slate of officers was the same as that elected in Ronneby, with the addition of F. Gerritzen, the leader of the conference in Utrecht, as another board member.[2]

The membership of the new Society had grown quickly since its formation. The report presented by the council to the membership at this 1939 meeting identified 68 members, and a footnote inserted before the publication of the proceedings in 1940 indicated that several new members had joined after the meeting. The total number identified by name in the published proceedings is 78, almost half of whom were Swedes (37), almost a sixth Germans (12), and the rest from other countries (5 Dutch, 5 American, 4 Swiss, 3 Norwegian, 3 Finnish, 2 Danish, and one member each from Algeria, Argentina, Italy, Canada, Turkey, England, and the Soviet Union).[3] Of the five members listed from the United States, John Welsh from Harvard was probably known in Europe, owing to his research on crustacean rhythms, and perhaps also T. H.

---

[1] "Vorstand der Internationalen Gesellschaft für biologische Rhythmusforschung, gewählt in Utrecht 1939," in "Verhandlungen der Zweiten Konferenz," ed. Holmgren, p. 7: President, Erik Forsgren (Sweden); Vice-President, Jakob Möllerström (Sweden); Secretary and Treasurer, Hjalmar Holmgren (Sweden); "Vorstandsmitglied" Arthur Jores (Germany); and F. Gerritzen (Holland); "Supplant" Antonia Kleinhoonte (Holland) and Hans Kalmus (England); "Revisor" Måns Arborelius (Sweden) and Ragnar Nilsson (Sweden).

[2] "In Ronneby 1937 gewählter Vorstand der Gesellschaft," p. 9.

[3] Holmgren, ed., "Verhandlungen der Zweiten Konferenz," pp. 11–14 ("Mitgliederverzeichnis") and pp. 15–18 ("Vorstandsbericht für die Jahre 1937–39"). It is interesting to note, in light of Jores's comment some years later about the Americans' lack of participation in early biological rhythm studies, that the only English member of the Society at this time was the ethnic German Hans Kalmus, who had left Prague. *Where were the Brits?* One of them, Kenneth Mellanby, from the Zoology Department of the University of Sheffield, contributed the paper "Rhythmic Activity in Domestic Insects" (pp. 89–98), so we know he attended the conference, but he is not listed among the members. We know there were researchers active in England, particularly at Oxford and on the Isle of Man, but they do not seem to have been in close academic contact with their European colleagues.

Bissonnette, who had published many articles, including one on photoperiodicity.[4] William F. Petersen from Chicago was broadly interested in correlating the cyclical nature of all cosmic occurrences and not engaged in experimental scientific research specifically in biological rhythms, but his work is nevertheless referenced in the European literature. The same can be said of the Soviet member, Alexander Chizhevskii (spelled Tchijevsky in the list).

Clearly, the Swedish members of the new Society dominated both its membership and leadership, and this dominance was also manifest financially. Whereas the proceedings of the 1937 meeting were later published in two issues of the German professional medical journal *Deutsche Medizinische Wochenschrifft* at the authors' expense and without mention of the *Societas*, its leaders (no doubt mainly President Forsgren and Vice President Möllerström) had now secured a gift from Sweden's Consul-General, Hugo Duhs and his wife, to cover the costs of publishing the proceeds of the Utrecht meeting as a special issue of the Swedish-dominated medical journal, *Acta Medica Scandinavica*. Control of the new Society was thus even more closely tied to Stockholm and *de facto* to medicine.[5]

The 1939 meeting was formally opened by Erik Forsgren, who began by reporting on the foundation of the Society and its interdisciplinary nature:

> In August two years ago a small flock of Rhythm enthusiasts, representatives of botany, medicine, and zoology, gathered in Ronneby (Sweden) at the first international conference on biological rhythm research. Today we are pleased

---

[4]Thomas Hume Bissonnette, "Sexual Photoperiodicity," *Journal of Heredity* 27 (1936): 171–80. John H. Welsh, "Diurnal Rhythm of the Distal Pigment Cells in the Eyes of Certain Crustaceans," *Proceedings of the National Academy of Sciences of the Unites States of America* 16 (1930) 386–95, reflects the author's studies in Cuba in 1929, which showed clear evidence of diurnal rhythms independent of natural stimulus by the light/dark cycle. Welsh identified his work as pioneering and placed it in the context of biological rhythms research, p. 388: "Animals kept in the dark showed no movement of the distal pigment cells, these cells constantly remaining in the position characteristic for the dark. . . . [This is] unquestionably due to a diurnal rhythm and such a case, as far as is known, has not previously been reported for eye pigment in crustaceans. It may be compared to the diurnal rhythm observed by Gamble and Keeble (1900) . . . by Slome and Hogben (1929) . . . (Easterly 1927) . . . (Moore, 1909) . . . (Crozier, 1920)." He published several more articles in the 1930s, leading up to his competent review of work in the field published the year after the Ronneby Brunn conference, John H. Welsh, "Diurnal Rhythms," *The Quarterly Review of Biology* 13 (1938): 123–39.

[5]Judging by the title, *Acta Medica Scandinavica* represented the Scandinavian medical community and not just the Swedish. However, it was a creature of the Karolinska Institute and published in Stockholm under chief editor Israel Holmgren (1871–1961), a professor at the Karolinska. The journal was clearly intended for an international audience, inasmuch as contributions would only be printed in English, French, or German. The importance of the Duhs's contribution is evident from the treasurer's report in Holmgren, ed., "Witschaftliche Stellung der Gesellschaft am 24. August 1939," published with the proceedings in Holmgren, ed., "Verhandlungen der Zweiten Konferenz," p. 19. The Society had collected 45 Swedish Crowns in 1938 and had a balance of 31.75 going into 1939, as individual members had paid for the publication of the 1937 proceedings. Contributions in 1939 amounted to 5,126.28 Crowns, owing mostly to Duhs's contribution of 5,000, leaving the Society with a balance of 4,960 after expenditures. From this, the publication of the 1939 proceedings was expected to use about 3000. Clearly the Society was not economically self-sufficient and dependent on donors' largesse.

that so many distinguished scholars have accepted our invitation and complied, that besides the already mentioned sciences, mathematics and now meteorology are represented.[6]

The creation of a new Society had clearly been warranted, he argued, both by the demand that was evident in the current turnout for the second meeting and by the special nature of the subject matter. Here he prioritized what for him was especially significant, namely, the medical importance of understanding biological rhythms, including the need to take them into consideration for both therapeutic and hygienic practice.[7] Given Forsgren's training and the nature of his research, a focus on rhythms with regard to human physiology is understandable, but he was also articulating the importance of a more general scientific study: "Plants and animals are often better than humans for more in-depth rhythm study. Rhythm research is also particularly suited to be a connecting link between the various disciplines of science."[8] He closed with a comment that was no doubt meant to dispel any perceived scholarly disagreements and encourage a spirit of collegial cooperation, but which also eerily reminds the modern reader of the political tensions that were palpable in Europe on the eve of World War II and the relative helplessness of academic scientists and clinicians in the grip of forces too large to comprehend or control:

> Rhythm is primarily a cosmic phenomenon. From a cosmic perspective, we are all quite small, as also are our temporary disagreements, which therefore

---

[6] Erik Forsgren, "Eröffnungsrede," pp. 23–25, in "Verhandlungen der Zweiten Konferenz," ed. Holmgren, p. 23: "Im August vor zwei Jahren versammelte sich in Ronneby (Schweden) eine kleine Schar von Rhythmusentusiasten, Vertreter für Botanik, Medizin und Zoologi zu der ersten, internationalen Konferenz für biologische Rhythmusforschung. Heute sind wir froh, dass so viele und hervorragende Gelehrte unsere Einladung Folge geleistet haben, dass ausser den schon genannten Wissenschaften jetzt auch die Mathematik und die Meteorologie repräsentiert sind."

[7] Ibid., pp. 23–24: "Es gibt wohl schon ziemlich viele internationale Gesellschaften und ein Skeptiker könnte vielleicht fragen, ob es motiviert ist, eine solche speziell für die Rhythmusforschung zu stiften. Ich meine, dass der lebhafte Anschluss zu dieser Konferenz diese Frage mit "ja" beantwortet. Dazu will ich in Kürze noch einige Beweggründe nennen. Für mich als Arzt liegt es nahe hervorzuheben, dass die Rhythmusforschung von einer sehr grossen Bedeutung für die praktische Medizin ist. Bei der *diagnostischen* Tätigkeit ist es zum Beispiel wichtig zu wissen, dass die Körpertemperatur, der Stoffwechsel die Zusammensetzung des Blutes, die Diurese u.s.w. Tageschwankungen rythmischen Natur unterworfen sind. Hinsichts der *Hygiene* spielen besonders die exogenen Rhythmusstörungen eine grosse Rolle. . . . Die *Therapie* muss dem Rhythmus angepasst werden." [There are already quite a number of international societies, and a skeptic could perhaps ask whether establishing one especially for rhythm research is warranted. I believe that the vibrant participation at this conference answers that question with a "yes." I want briefly to name a few additional reasons. For me as a doctor, it is obvious to emphasize that rhythms research is of very great importance for practical medicine. In *diagnostic* work it is important to know, for example, that the body temperature, metabolism, the composition of blood, diuresis, etc. are subject to daily fluctuations that are rhythmic in nature. In terms of hygiene, exogenous rhythm disturbances play an especially important role. . . . *Therapy* must be adapted to the rhythm.] In hindsight, one can wonder at Forsgren's use of the term "Anschluss" at this particular historical moment.

[8] Ibid., pp. 24–25: "Die Pflanzen und die Tiere eignen sich oft besser als der Mensch für mehr eingehende Rhythmusstudien. Die Rhythmusforschung ist auch besonders dazu geeignet ein verbindendes Glied zwischen den verschiedenen Disziplinen der Naturwissenschaft zu werden."

need not disturb our friendship. I cherish the hope and conviction that the forthcoming consultation will lead to a friendly, informative and beneficial exchange of views.[9]

The photograph that accompanies the published proceedings of the 1939 conference shows 32 people, at least ten of whom are women. Assuming that the seventeen papers published in the proceedings reflect the oral presentations at the meeting, along with the opening speech and concluding "Schlusswort" by Forsgren, there must have been a dozen or more spouses or other ancillary participants in attendance, seven of whom are identified in the photograph (Figure 7.1).[10] The seventeen papers were presented by fifteen researchers; six from Germany, four from the Netherlands, four from Sweden, and one from England. Sessions were chaired by presenters, except for Antonia Kleinhoonte, who presided over one session but did not contribute a paper. It might be expected that the location in Utrecht would have encouraged participation from the Netherlands and western Germany in particular, and indeed, with the exception of Dr. Philipsborn from Obersdorf in Allgäu, southern Bavaria, and Dr. Menzel from Tübingen, the German participants came from the west: Hamburg, Frankfurt am Main, and Köln. The Swedes who presented were those who had been at the first meeting and were part of the Society's administration, with the exception of Yngve Edlund, who presented jointly with Hjalmar Holmgren and later collaborated with him on an article on liver function.[11] Clearly the Swedes were still an important component both in the leadership of the new Society and in participation, although shifting the meeting to a continental location had brought more researchers into the organization and broadened its scope.

Germans in particular had been interested in biological rhythms since the late-nineteenth-century work of Sachs and Pfeffer and the follow-up research and discussion that Pfeffer's plant physiology generated, and one might well expect them to have been major contributors to the new field as it became organized, had not the problematic political conditions within Germany in the late 1930s made it unattractive for the Germans to dominate what was meant to be an international society. Moreover, the German scientific community was

---

[9] Ibid., p. 25: "Die Rhythmus ist vor allem eine kosmische Erscheinung. Aus kosmischem Gesichtspunkt sind wir alle ziemlich klein, wie auch unsere jeweiligen meinungsverschiedenheiten, die darum unsere Freundschaft nicht stören brauchen. Ich hege darum die Hoffnung und Überzeugung, dass die bevorstehende Beratung einen freundschaftlichen, lehr- und segensreichen Meinungsaustausch zu Folge haben wird."

[10] Persons identified in the photograph who were neither presenters nor members of the Society's administration are J. W. Langelaan (professor at Baarn, Holland), H. Laurell (physician in Halmstad, Sweden), and Bernhard de Rudder (professor at Frankfurt am Main)—listed as members—and Jeslings, Banky, de Langen, and Koumans.

[11] Y. Edlund and Hj. Holmgren, "The Rythmical Variations of the Liver Glycogen and the Pyruvic Acid of the Blood in Experimental Obstructive Jaundice," *Acta Medica Scandinavica* 120 (1945): 107–29.

Figure 7.1 Participants in the second meeting of the *Societas pro studio rythmi biologici*, Utrecht 25–26 August 1939. Photograph published at the beginning of "Verhandlungen der Zweiten Konferenz der Internationalen Gesellschaft für Biologische Rhythmusforschung am 25. und 26. August 1939 Utrecht (Holland)," ed. Hjalmar Holmgren, *Acta Medica Scandinavica* Suppl. 108 (1940). Reproduced with permission of the current publisher of this journal, John Wiley & Sons, Ltd.

itself fractured by Nazi politics, and Jores and Kalmus, founding members of the Society, had been excluded from the ranks of German academics. Nevertheless, the Society's council report at the 24 August 1939 meeting announced that Arthur Jores had proposed to create a separate German society for biological rhythms research, which he envisioned as an affiliate institution of the International Society; he proposed that the *Societas* create two libraries, one to serve the German circle in Hamburg, to be administered by Jores, and one in Stockholm, to be administered by Möllerström.[12]

Germans had been prominent in the field for decades, so it is not surprising that they thought they should have their own affiliate society; nor is it therefore surprising that Germany was chosen as the location of the next meeting of the *Societas*, to be held at a time and place to be determined later.[13] Given Jores's salience in the Society's council from the beginning and Rose Stoppel's long career as a rhythms researcher, both of them at Hamburg, that well-connected Hansa city would have been the likely choice for the 1941 meeting. But, as a consequence of the events set in motion later that week, voluntary cooperation between Germany and much of the rest of the world was not a possibility for the next five years, and the meeting planned for Germany did not materialize, at least not in Germany.[14]

---

[12] "Protokoll der Vorstandssitzung in Lunteren am 24. August 1939 (Holland)," in "Verhandlungen der Zweiten Konferenz," ed. Holmgren, p. 20: "Auf Vorschlag von Dr. A. Jores wurde beschlossen, in Deutchland eine Vereinigung für biologische Rhythmusforschung zu bilden. Diese Vereinigung soll als Ganzes einen Bestandteil der Internationalen Gesellschaft bilden. Die Mitglieder sollen ihre Beiträge an den Vorstand der deutschen Vereinigung erlegen. Ferner wurde die Erreichtung von zwei Bibliotheken beschlossen, eine für den deutschen Kreis in Hamburg, Vorsteher Dr. Jores, und eine in Stockholm, Vorsteher Dr. Möllerström." [At the suggestion of Dr. A. Jores it was decided to form an association for biological rhythm research in Germany. This association will form a part of the greater International Society. The members should give their contributions to the board of the German association. It was also decided that two libraries be established, one for the circle in Hamburg, Germany, headed by Dr. Jores, and one in Stockholm, headed by Dr. Möllerström.]

[13] Ibid., p. 20: "Schliesslich wurde vom Vorstand beschlossen, den nächsten Kongress 1941 in Deutschland abzuhalten. Zeit und Ort werden später mitgeteilt werden." [Finally, the council decided to hold the next congress in 1941 in Germany. The time and place will be communicated later.]

[14] Erwin Bünning's absence from the 1939 meeting at Utrecht can be explained by his recent return to Germany from Indonesia and subsequent conscription into the Wehrmacht, but his absence from the list of members at the end of 1940 and failure to participate in the 1937 meeting is somewhat puzzling, given the prominent place he would claim in the development of biological rhythms research later in his life. He was publishing already in the early 1930s and was at least intellectually connected with Rose Stoppel's work. Perhaps he was not yet in a professional position that could support his membership and attendance at foreign meetings, or perhaps he was not getting on well with Stoppel, Jores, Kalmus, and other German members of the Society. Erwin Bünning, "Fifty years of research in the wake of Wilhelm Pfeffer," *Annual Review of Plant Physiology* 28 (1977): 1–22, does little to explain this. He notes that he and Kurt Stern were engaged as postdoctoral fellows at Frankfurt in 1928 and soon discovered that red light triggered plant behavior, that he obtained an assistantship at Jena in 1930, but was forced to leave after two years on account of his suspected Communist sympathies, but then says little of his activities in Germany between 1933 and the end of the war.

Between the close of the second meeting of the Society at Utrecht and the planned third meeting, which should have taken place in late summer 1941, if it had followed the alternate-year trajectory established by the first two, Germany had not only initiated war with England and France by invading and annexing western Poland in the fall of 1939, but also overrun the low countries and occupied or subdued France. As part of Hitler's strategy to secure western Europe and protect his northern flank, the German military machine also invaded Denmark and Norway on 9 April 1940. Denmark capitulated immediately, and Norway followed two months later, after a bitter and confusing struggle. Sweden— isolated and caught between its two closest Scandinavian neighbors, which were occupied by Germany, and its other Nordic neighbor, Finland, which was partly ethnically Swedish and allied with Germany against the Soviet Union—opted for formal political neutrality. Under these circumstances, the leaders of the *Societas*, predominantly Swedes, chose to convene a conference in Stockholm at about the time one should have been scheduled for Germany. Not surprisingly, it was attended mainly, perhaps exclusively, by Swedes.[15]

I have not yet located a program for this meeting or archival material connected with its organization, but it is evident from Hjalmar Holmgren's preface to the proceedings, which were published in the Stockholm Medical Association's journal in the month following the meeting, that the Society's leaders understood it to be the third meeting of the International Society at the time they convened it.[16] Holmgren's preface was clearly intended to explain to the journal's readership, members of the Stockholm Medical Association and therefore mainly professional physicians, why a series of articles on biological rhythms should be published in their journal. To this end he briefly recounted the formation of the Society in Ronneby in 1937, which he duly identified

---

[15] I have not yet located any attendance sheet or other archival matters pertaining to this meeting, but the published proceedings are entirely by ethnic Swedes and the publication is in Swedish, suggesting that there was little international about the meeting. Travelers from Allied countries would have had to traverse Axis-controlled airspace, ruling out their participation at the meeting. Because Sweden was formally neutral, and Germans were transporting supplies through Swedish waters to Finland and shipping Swedish iron ore out of Kiruna by way of the railroad to Narvik in occupied Norway, the presence of German soldiers on Swedish soil was a sensitive issue, and this may have discouraged German scientists from visiting Sweden, too, even if they would have had the means or permission to leave greater Germany.

[16] Hjalmar Holmgren, "Förord till en artikelserie i MFT," *Medicinska Föreningens Tidskrift* 19 (1941): 199–200, p. 199: "I de följande numren av Medicinska Föreningens Tidskrift kommer genom redaktionens vänliga tillmötesgående att publiceras en serie artiklar, vilka hällets såsom föredrag vid Internationella sällskapets för biologisk rytmforskning sammanträde i Stockholm den 22 augusti 1941." [In the following issues of *Medicinska Föreningens Tidskrift* comes, through the editor's friendly willingness to publish, a series of articles which were held as lectures at the meeting of the International Society for Biological Rhythm Research in Stockholm, 22 August 1941.]

by its Latin title *Societas pro Studio rythmi biologici* rather than the German title that had graced the previous published proceedings, and explained that its purpose was to bring together researchers from various disciplines who were working on biological rhythms, "which are of extraordinary significance for all living things."[17]

Because the Stockholm meeting is little acknowledged in the literature, and its proceedings were published in Swedish, a fuller consideration is warranted here. The 1941 meeting was a one-day affair, convened Friday, 22 August. In the absence of a program, relevant correspondence, or other indications, we might suppose that the meeting took place at the Karolinska Institute or perhaps at the Wenner-Gren Institute, because these would have been natural venues for the organizers, but possibly also at the Stockholm Medical Society's meeting house or even at Stockholms Högskola.[18] We can reconstruct the presentations at the meeting from the proceedings published in the Stockholm Medical Society's journal. Not including Hjalmar Holmgren's foreword to these papers, which would not have made sense as an opening of the meeting, there are five published papers that are each identified by footnote as presentations given at the 22 August 1941 meeting of the *International* Society for the Study of Biological Rhythm ("Föredrag vid Internationella Sällskapets för biologisk rytmforskning sammanträde"), plus a final paper that is not thus identified, by Holmgren himself.[19] There is no indication that the papers were presented in the order they are printed. The first paper is by Erik Forsgren and reports results from his research on the rhythmical characteristics of certain liver and stomach functions, with which he had well over a decade's experience. He began by pointing out that his discovery of the secretory and assimilatory phases of the liver's cells provided a "morphological" (i.e., anatomical) basis for rhythmicity in liver function, and argued that, given the liver's importance for the body's metabolism in general, rhythms in other bodily functions follow logically from this basic cycle of assimilation and

[17] Ibid., pp. 199–200: "Sällskapet har bl. a. satt som sitt mål att söka sammanföra forskare från skilda områden kring de rytmiska problemen samt arbeta på ett ökat intresse för dessa problem, vilka äro av utomordentlig betydelse för allt levande." [The Society has among other things set its goal to seek to bring together researchers from different areas around the rhythmic problems and work for an increased interest in these problems, which are of extraordinary significance for all living things.] Holmgren's evocation of the Latin title for the Society was likely intended both to identify it as truly international and to put some distance between the Swedes and the Germans at that particularly sensitive moment in Scandinavian history.

[18] Eventual search of archival materials at the Wenner-Gren Institute and the Karolinska Institute may yet shed further light on the planning and staging of this meeting.

[19] The papers appear serially in *Medicinska Föreningens Tidskrift* 19 (1941): 199–210; 227–47, and 20 (1942): 2–15. The first footnote to each of these papers reads "Föredrag vid Internationella Sällskapets för biologisk rytmforskning sammanträde den 22 aug. 1941" [Lecture at the meeting of the International Society for Biological Rhythm Research 22 August, 1941], with the exception of Romell's, which indicates that certain changes were made to the original presentation.

secretion of glycogen.[20] Drawing on his broad knowledge of developments in the field generally, Forsgren sketched out a connection between light and body rhythms that would be a major subject of research over the next fifty years: "Observations on plants and animals show that metabolism is affected by sunlight, which stimulates the functioning of the entire individual. The liver is certainly protected from the direct effect of the sun, but it can be affected indirectly through organs that sense radiation, via the central nervous system and the hormones."[21] Pointing to his research on rhythmic changes apparent with diabetes, which revealed instances of significantly delayed peaks in the excretion of sodium chloride in the urine, Forsgren noted the connection between variability in rhythms and abnormal or pathological functions and speculated on the effects of artificial illumination on human rhythms.[22] He concluded his paper by pointing out that knowledge of the rhythmicity of human vital processes was still quite incomplete, but that what little was known was clearly of great significance for both preventative medicine and for therapeutics. Many questions remained, he wrote, not least of which was the relationship between the innate and individual endogenous rhythms and exogenous or external factors, such as illumination, eating, long-distance travel, and physical exercise in determining biological rhythms.[23]

The second paper was given by Ansgar Roth, picking up a more cosmic theme that goes back to Arrhenius and Chizhevskii. He took as his point of departure a quotation from Ludwig Klages that Jores had alluded to in his introduction to the papers from the Ronneby Brunn meeting and that one

---

[20] Erik Forsgren, "Om tidsinställning av några funktioner, spec. leverns och ventrikelns hos människan," *Medicinska Föreningens Tidskrift* 19 (1941): 200–210, p. 201: "Dygnsvariationerna hos den inre ämnesomsätt-ningen, kroppstemperaturen, diuresen etc. kunna antagas i första hand vara baserade på leverns rytmiska funktion." [Diurnal variations in the internal metabolism, body temperature, diuresis, etc. can be assumed in the first instance to be based on the liver's rhythmic function.]

[21] Ibid., pp. 201–2: "Observationer på växter och djur tyder på att ämnesomsättningen påverkas av solljuset, som stimulerar verksamheten hos hela individen. Levern är visserligen skyddad mot direkt solpåver-kan men den kan påverkas indirekt genom strålningskänsliga sinnesorgan via centrala nervsystemet och hormonerna."

[22] Ibid., p. 203: "Som orsak till en mer eller mindre vanlig försening av ämnesomsättningsrytmens tidsinställning få vi i första hand tänka på det artificiella ljuset. . . . Den icke naturenliga belysningen användes sannolikt av flertalet människor till att förlänga dagen på kvällsidan, vilket skulle kunna förklara en mer eller mindre allmän försening av ämnesomsättningens tidinställning." [As the cause of a more or less normal delay of the timing of the metabolic rhythm we primarily think of the artificial light. . . . Non-naturally compatible lighting is used probably by most people to extend the day in the evening, which could explain a more or less general delay of the timing of metabolism."

[23] Ibid., p. 205: "Redan det lilla vi vet är av stor betydelse för vårt profylaktiska och terapeutiska handlanda. Men de återstår många frågor att besvara, varpå som exempel kunna nämnas följande. I vilken mån är grundrytmens tidsinställning *"endogen"* (medfödd, individuellt bunden) eller *"exogen"* (påverkbar av yttre faktorer såsom belysning, näringstillförsel, geografisk förflyttning, träning etc.)." [Already the little we know is of great importance to our prophylactic and therapeutic treatment. But there are still many questions to answer, for example, the following: To what extent is the timing of the fundamental rhythm "endogenous" (congenital, individually bound) or "exogenous" (influenced by external factors such as lighting, nutrition, geographic mobility, training, etc.)?]

finds mentioned by other writers on biological rhythms. This concerned the difference between biological rhythms and mechanical rhythms, a point of philosophical importance in discussions about whether biological functions can be reduced to mechanical properties (vitalism vs. mechanical determinism):

> Ludwig Klages has defined the concept of rhythm in contrast to cycle and formulated the distinction as follows: "Rhythm is the recurrence of similar in similar time intervals, cycle is the recurrence of like in like intervals." Rhythmic phenomena occur everywhere in nature; cycle, however, is a human invention, suitable for mechanical use.[24]

Clearly the philosophical and social consequences of reductionism were still cogent in the scientific community and were now producing medical concerns about shift work and industrial stress as well. But Roth then elaborated on various kinds of cosmic cycles, from terrestrial rotation, to the pulsation of variable (binary) stars and to the regularity evident in sunspot cycles, and the thrust of his argument is to call into question whether there is in fact a meaningful distinction between the precision of human machines and the machinery of the cosmos, between cycle and rhythm.[25]

Måns Arborelius's paper considers diurnal rhythms and states of health. His research had shown him that a useful means of assessing biological rhythm in humans is to assay the urine for sodium chloride, which under normal conditions exhibits a daily rhythm with a minimum in the evening and a maximum in the afternoon.[26] To facilitate comparison of normal and pathological sodium excretion rhythms he created what he called a "rhythm number" (*rytmtalet*), which is the ratio of the afternoon maximum to the evening minimum urine salt concentration. This number is normally above three, but he defined it in such a way that a patient exhibiting an inverted rhythm, with the maximum in the night and the minimum during the day, has a rhythm number less than one.[27] Arborelius recognized that any sickness that affected

---

[24] Ansgar Roth, "Rytmiska företeelser I universum," *Medicinska Föreningens Tidskrift* 19 (1941): 205–10, p. 205: "Ludwig Klages har preciserat begreppet rytm i motsats till takt och formulerat skillnaden på följande sätt: 'Rhythmus ist die Wiederkehr von *Ähnlichem* in *ähnlichen* Zeitabständen, Takt ist Wiederkehr von *Gleichem* in *gleichen* Zeitabständen'. Rytmiska företeelser förekomma överallt i naturen, takt däremot är ett människopåfund, lämpat för maskinellt bruk."

[25] Ibid., p. 205: "Denna definition kan vara lika bra som någon annan, om man blott kommer i håg att någon sträng gräns mellan de bägge begreppen inte kan upprätthållas. Jorden roterar kring sin axel med samma precision som den mest fulländade motor." [This definition can be as good as any other, if only one remembers that no strict line between the two concepts can be maintained. The earth rotates on its axis with the same precision as the most perfect engine.]

[26] Måns Arborelius, "Om dygnsrytmen vid sjukdomstillstånd," *Medicinska Föreningens Tidskrift* 19 (1941): 227–33, p. 227: "Normalt sker denna koksaltutsöndring uttalat rytmiskt med ett minimum under förnatten och ett utsöndringsmaximum på eftermiddagen." [Normally this salt excretion is expressed rhythmically with a minimum during the evening and a secretion peak in the afternoon.]

[27] Ibid., p. 228.

the kidney function would also likely affect salt excretion, but even so thought that the device was useful. He tracked the rhythm number in diseases characterized by fevers, defects in circulation, malignant tumors, gastrointestinal disorders, liver diseases, kidney diseases, and damage caused by alcoholism. He concluded by defining rhythmic disorders "as a more independent disease concept":

> Rhythm disorder can occur as a relatively independent disease. I have had the opportunity to make further observations of such cases. In particular there is a large community for which the appearance of arrhythmia ought to be noted, namely in all shift workers, where the relatively frequent changing of the shift time is likely to favor the emergence of arrhythmia with particular nervous symptoms, which would otherwise be difficult to interpret.[28]

Arborelius's experience treating nightworkers and others with shifted schedules pointed the way to what would become a major concern of biological rhythms researchers.

Jakob Möllerström's paper on periodicities evident in the excretion of derivatives of pyridin with diabetic patients is of a much more biochemically technical nature than are the other papers, drawing on his strength as a metabolic researcher at the Wenner-Gren Institute. Pyridin is implicated in the fundamental metabolic processes of the body, and Möllerström set out to study it by conducting urine tests under varying conditions of carbohydrate metabolism. For this purpose, hospitalized diabetic patients, who were already being held on strictly enforced and differing carbohydrate diets, were particularly suitable and convenient research subjects, especially for somebody who had worked at St. Erik's Hospital, where a number of diabetic patients were treated. Möllerström had been testing the urine of these patients at regular time intervals over the years, so this was not a new methodology, but rather an application of an established procedure to assay a new component of the urine. The bulk of this rather lengthy paper comprises an account of the clinical case studies of seventeen patients, some taking insulin and some not. Each is identified by first name, initial of surname, and date of birth and is accompanied by a brief case history. For each case Möllerström presents a chart of excreted pyridin, glucose, β-oxybutyric acid, and pyruvic acid, the rhythms of which he noted in the case histories. The dates on the charts indicate that the studies were conducted between mid-May and mid-June 1941, just months before the conference, and owing to the freshness of the study and desirability of

---

[28] Ibid., pp. 232–33: "Rytmrubbningen kan alltså uppträda som en tämligen självständig sjukdom. Jag har haft tillfälle att göra ytterligare iakttagelser över sådana fall. Speciellt är det en stor befolkningsgrupp, där uppträdandet av rytmrubbning bör uppmärksammas, nämligen hos alla skiftesarbetande, där det förhållandevis täta ombytet av skiftestiden är ägnat att gynna uppkomsten av rytmrubbning med framför allt nervösa symptom, som annars bli svåra att tolka."

additional data before reaching conclusions, Möllerström declined theoretical speculation. Nevertheless, he could not refrain from arguing that the rhythmicity of the metabolic products pointed to a deep-seated endogenous rhythm in oxydation reactions, which required further research, and that the diabetes study offered an opportunity to investigate these processes, which were otherwise inaccessible to experimental research in humans.[29]

In contrast to Möllerström's paper, which delivers fresh, partly digested experimental results, Gunnar Romell's paper is more like a review article, summarizing decades of research in plant rhythms. Romell was apparently not in attendance at the 1937 meeting at Ronneby Brunn or the meeting in Utrecht, which suggests that there might have been some local political or disciplinary distinctions separating rhythms enthusiasts in Stockholm during the formative years of the *Societas*. But the inclusion of his lengthy, synthetic paper on plant rhythms in the 1941 meeting, along with Roth's paper on cosmic rhythms, indicates that the Swedes were at least by then not viewing rhythms research in a strictly medical framework. Indeed, Romell's paper places the work of the *Societas* in a broad, multinational research context with clear roots in the early German, Dutch, and also Swedish investigations of biological rhythms.

Romell began his paper with Darwin's observations on plant movements and subsequent physiological studies of diurnal patterns in reproduction, transpiration, assimilation, osmotic pressure, tissue permeability, acidity, growth, cell division, and nuclear division. Then he noted studies that have shown the presence of what we today call "ultradian" rhythms, namely, those with periods significantly shorter than 24 hours ($\tau < 24$ hr) , citing the work of his Swedish colleague Martin Stålfelt on cell rhythms in leaves, and then also phenomena that exhibit "infradian" ($\tau > 24$ hr) rhythms, exemplary in the reproduction of the palolo worm and some other sea animals that have monthly or bimonthly periodic behaviors and in the foliage shifts of tropical trees that replace their leaves between two and six times per year. Romell pointed out that these rhythmicities fall into two clear groups; those with periods that correspond to environmental rhythms (daily, monthly, yearly) and those that

---

[29] Ibid., pp. 246–47: "Jag vill icke ingå några teoretiska spekulationer eller resonemang . . . Jag vill endast konstatera det intressanta faktum, att den verksamma principen i ett av våre dehydrerande andingsferment i stor utsträckning visar en dygnsperiodisk utsöndring . . . så ligger uppenbarligen den endogena rytmiken och ämnesomsättningens periodiska förlopp djupt förankrad i de biologiska oxidationsprocesserna. . . . Vi ha i diabetesstudiet en möjlighet att tränga dessa problem närmare, vilka eljest äro i stor utsträckning otillgängliga för experimentell forskning." [I will not enter into any theoretical speculation or reasoning . . . I will only note the interesting fact that the active principle in one of our dehydrating respiratory ferments largely shows a diurnal periodic secretion. . . . Thus apparently the endogenous rhythms and the metabolism's periodic progress is deeply rooted in the biological oxidation processes. . . . We have in the diabetes study an opportunity to penetrate these problems more closely, which otherwise are largely inaccessible to experimental research.]

do not. For the former group it is difficult to discern whether observed rhythms are endogenous or induced by the environmental rhythms because they generally coincide with environmental triggers. He stated that the latter must be the result of endogenous rhythms, which bear comparison to the oscillations of a pendulum, except that they are affected by temperature.[30] This is an important claim, inasmuch as the temperature dependency or independency of biological rhythms was already under scrutiny and would become a key element in arguments about the endogenicity or exogenicity of the causes of rhythms. Romell was not the first to draw attention to this, but he was clearly identifying a consideration that would remain in the foreground of biological rhythms research for at least the next three decades.

Turning to diurnal rhythms, Romell recounted a long history of observations and experiments on plant sleep and movements, beginning with Linnæus's observation that individual flowers opened and closed at distinct times, such that one could use them collectively as a kind of clock, but also recounting experimental studies by De Candolle, Dutrochet, and especially by Pfeffer and his commentators—notably Semon, Karsten, Stoppel, Stålfelt, and Kleinhoonte. Stålfelt was Romell's colleague in Stockholm, and Stoppel and Kleinhoonte were involved with the Society, so his recognition of their work is unsurprising. Of interest is Romell's keen awareness of the work of Erwin Bünning in Germany, who had not participated in either of the first two meetings of the Society and was not a member. Romell recounted Bünning's revelation that Stoppel's observations on leaf movements were vitiated by her use of a red-orange "safelight" in the laboratory, but still harbored some suspicions that researchers had not ruled out unknown exogenous factors that might be covertly synchronizing movements that were interpreted by Bünning and others to be the result of endogenous causes.[31] Frank A. Brown, Jr. adopted this cautionary perspective in the 1950s, and his insistence on the methodological point that experimenters had not excluded their subjects from *all* possible external timing stimuli became a key point of contention and divided the rhythms research community for the next couple of decades. Romell, however,

---

[30] Lars-Gunnar Romell, "Något om rytm och periodism hos växter," *Medicinska Föreningens Tidskrift* 20 (1942): 2–10, p. 3: "De sistnämnda rytmerna framstå omedelbart såsom egenrytmer jämförbara med en pendels svängningar, helst som deras hastighet brukar påverkas av temperaturen." [The latter rhythms appear immediately as characteristic rhythms comparable to a pendulum's swings, especially as their speed is usually influenced by temperature.] The footnote to this sentence reads "Det är t. ex. endast vid 30°–35° C som *Desmodium gyrans* svänger med sidosmåbladen så fort att man kan se det. Vis högra och lägre temperatur gå rörelserna långsammare." [It is, for example, only at 30°– 35° C that *Desmodium gyrans* swings its small side leaves so fast that one can see it. If higher and lower temperatures, the movements go more slowly.] This comment suggests that Romell himself might have been doing such experiments.
[31] Ibid., p. 8: "Även i de ännu helt ouppklarade fallen (kärndelningsrytmen och blödningsrytmen) kan de gott hända, att någon ännu okänd faktor är med i spelet." [Even in the still completely unsolved cases (nuclear division rhythm and bleeding rhythm), it can well be that some yet unknown factor is in play.]

also acknowledged that Bünning's experiments had revealed that some plants seem to have characteristic rhythms that were heritable, strongly suggesting that timings were somehow inherent in the structures of the plants, namely, the rhythms were endogenous.[32]

Romell's paper indicates that he found the key question facing biological rhythms researchers to be the endogenous or exogenous nature of rhythmicity and its causes. The observation that would prove particularly telling, he thought, was whether there was a connection between diurnal rhythms and annual photoperiodic rhythms, which Bünning's latest research suggested.[33] Such a result, he wrote, would have far-reaching significance, because it would suggest commonalities between animal and plant behaviors and, consequently, a common ontological (structural) basis. This was a conclusion with metaphysical implications:

> Should it be confirmed that it is the endogenous diurnal rhythm that underlies and determines plants' photoperiodic adjustment and reactions, then this is a matter of very great interest from various points of view. Among other things, it seems to me to indicate that there is not only a superficial similarity between plants' and animals' or our own sleep—as the scientists have tended to be so eager to point out—but that in both cases it is a question of a need for a repair phase with a well defined length, returning in a fairly definite rhythm, or of a corresponding rhythmic alternation between different activities. Linnæus thought something along this line when he said that "it has pleased the Almighty Creator, inasmuch as no living thing that lacks some form of rest period continues to exist, to grant most, if not all plants an analogue to sleep." . . . Recently Bünning (1935) has deduced interesting parallels between animal and plant endogenous diurnal rhythms. Finally, it might be that old Linnæus got it largely right with his innocent statement, which for a hundred

[32] Ibid., p. 8: "Det är nämligen så, att efter de senaste årens undersökningar växternas sömnrörelser nu igen (liksom under tiden före Pfeffer) i stor usträckning klart framstå såsom uttryck för en i växterna inneboende dygnsrytm, som fasreguleras men såsom rytm icke direkt betingas av den naturliga växlingen mellan dag och natt. . . . Individuella skillnader i periodlängd och svängningskurvans form kunna vara ärftliga (Bünning 1932 och 1937)." [The fact is, that after the recent year's surveys, the sleep movements of plants are again (just as before Pfeffer) to a great extent clearly seen as an expression of one of the diurnal rhythms inherent in the plants, which has a fixed phase, but as a rhythm is not directly determined by the natural alternation between day night. . . . Individual differences in period length and the shape of the movement curve [i.e., as it appears on kymograph charts] can be inherited (Bunning 1932 and 1937).]

[33] Ibid., p. 9: "De sist anförda erfarenheterna tyda på ett samband mellan fotoperiodisk inställning och endogen dygnsrytmik. Finns ett sådant samband, kan egenrytmens karaktär hos en växt uppenbarligen ha en mycket stor, ja avgörande betydelse ur urvalssynpunkt. Bünning (1937, 1939) menar, att saken förhåller sig så, och han har gjort försök för att pröva den tanken." [The last quoted experiences suggest a connection between photoperiodic adjustment and endogenous diurnal rhythm. If there is such a connection, then the natural frequency characteristic of a plant obviously can have a very large, even decisive importance from the point of selection. Bünning (1937, 1939) argues that this is the case and he has made attempts to test that idea.]

years or more has generally been regarded as an expression of the greatest naivete.[34]

Whether this was meant as a Christian teleologist's parting shot at reductionism in an age of Darwinian materialism, or simply a Swedish scientist's nationalist nod to his renowned forebear is open to interpretation.[35]

The final paper of the 1941 cohort is by Hjalmar Holmgren, who had also provided the foreword and *de facto* functioned as the editor of the proceedings. No doubt he also played an important role in hosting the meeting. Given the metaphysical note on which Romell's paper ended, and the fact that Holmgren's paper is not identified as one presented at the meeting (unlike the other five), it is tempting to see it as the editor's concluding comment on the subject of biological rhythms research, included here to bring some closure to the set. He chose to address the problem of endogenous and exogenous causes for biological rhythms, which was by then at least a half century old, but still not resolved satisfactorily (nor would it be for another two decades or more).[36] Holmgren reviewed briefly a wide range of rhythmic factors evincing connections between light rhythms and biological rhythms, rhythms of atmospheric electrical charge and biological rhythms, correlations between the rhythms of urobilin excretion in the urine and body temperature, rhythms in body temperature in newborn children, and glycogen rhythms in the livers of guinea

---

[34] Ibid., pp. 9–10: "Skulle de bekräftas, att det är den endogena dygnsrytmen som ligger under och bestämmer växternas fotoperiodiska inställning och reaktioner, så är dette en sak of mycket stort intresse ur olika synpunkter. Bland annat tycks det mig tyda på att det icke blott är en ytlig likhet mellan växternas och djurens eller vår egen sömn—såsom fackmännen ha brukat vara så angelägna om att påpeka—utan att det i bägge fallen är fråga om ett behov av en reparationsfas med någorlunda bestämd längd, återkommande i en någorlunda bestämd rytm, eller av en motsvarande rytmisk växling mellan olika aktiviteter. Något i den stilen trodde Linné, som menade, at 'det har behagat den allsmäktige Skaparen att, alldenstund intet levande, som saknar vilans omväxling, kan bliva bestående, tilldela de flesta, om icke alla växter ett analogon till sömnen'. . . . Nyligen ha intressanta paralleller mellan djurs och växters endogena dygnsrytmik dragits av Bünning (1935). Till sist blir det kanske så, att gamle Linné får ganska rätt med sitt troskyldiga uttalande, som under hundra år eller mer säkerligen allmänt har betraktats såsom ett uttryck för den största naivitet."

[35] It is instructive to compare Romell's sentiment with the near contemporary statement by J. Arthur Thomson, *Biology for Everyman*, ed. E. J. Holmyard, (New York: E.P. Dutton, 1935), p. 762: "Meanwhile, we welcome all that the chemist and the physicist can tell us about the ongoings in the living body; we remain none the less convinced that organisms are more than mechanisms; and we suggest as answer to our question, that Life is a dance of enchanted particles with Mind as the music."

[36] The existence of endogenous causes for rhythms and the role of exogenous "cosmic" factors continued to dog discussions within the field into the 1980s, especially the sometimes bitter exchanges between Frank A. Brown, Jr. and Colin Pittendrigh (and his partisans), although by then Brown was almost alone in insisting that the case for endogenous "timers" or causes for rhythms had not been proven. Nevertheless, as late as 2007 Peter Barlow assembled various drafts and material by Gunter Klein arguing that the experimental evidence presented by Erwin Bünning and others could be explained on the basis of rhythms in gravitational forces that none of the chronobiological experiments had successfully isolated. See Gunter Klein, *Farewell to the Internal Clock: A Contribution in the Field of Chronobiology* (New York: Springer Science+Business Media, 2007).

pigs and rats. From these and many other experiments Holmgren concluded that there is evidence for both endogenous and exogenous causes of rhythms and he evoked Jores's division of the causes of rhythms into (1) endogenous causes that are internal to an individual organism; (2) exogenous, cosmic causes that act directly on an organism; and (3) environmental causes that affect rhythm but cannot be considered cosmic, such as the timings of meals, rest periods, work periods, and the rhythmic behavior of other individuals or other species, which we might call social causes. There is evidence for all three kinds, Holmgren wrote, although experimental evidence for the social causation is not solid, inasmuch as the induction of an inverse rhythm by inverting the social causes, which would demonstrate the cause and effect relationship, has not been successfully accomplished.[37] The model that Holmgren preferred is not one of endogenous versus exogenous *causes*, but rather of endogenous, exogenous, and social *factors* that contribute variously to the production of rhythmic behaviors. He imagined that cosmic *factors* might act as triggering moments (utlösande moment) that synchronize fundamental organismal rhythms that are endogenous, and it follows that when organisms are isolated from such synchronizing moments they become arrhythmic and this gives rise to a pathogenic condition. However, Holmgren concluded that biological rhythms are complicated and that his findings should not be regarded as a final explanation. In short, Holmgren provided a kind of conclusion to the conference presentations, not by presenting any new material, but by conjuring up the diversity of experimental evidence for biological rhythms of various sorts and offering a logical explanation for how they might arise, leaving the reader with the understanding that sorting out the causal mechanisms for biological rhythmicities remained a central *desideratum* in the field of biological rhythms research.

---

[37] Hjalmar Holmgren, "Till frågan om de faktorer, vilka utlösa och utforma rytmen," *Medicinska Föreningens Tidskrift* 20 (1942):10–15, p. 15.

8

The Society Becomes
More International

With the Second World War over and some normalcy returning, at least to the West, the International Society for the Study of Biological Rhythm finally met in Germany. The meeting convened in Hamburg 30 September–1 October 1949, eight years after it was originally scheduled, under the slate of officers who were elected at Utrecht in 1939, namely, under the leadership of President Erik Forsgren and other founding members.[1] This meeting was identified in the published proceedings as the third meeting of the International Society; subsequent meetings followed this numeration, and, despite the fact that the Society's officers had regarded the 1941 meeting as its third international meeting, the 1941 Stockholm meeting was now *de facto* relegated to the status of national meeting.[2]

The published proceedings of the Hamburg conference comprise Jores's opening speech and thirty papers, three of which were coauthored. Altogether, the proceedings involved twenty German contributors, eight Swedes, one English (Kalmus, although he was ethnically German), and one Swiss, for a total of thirty. The photograph accompanying the proceedings shows thirty-one heads and identifies twenty-one of them, suggesting a good correspondence between the photograph and the contributors (Figure 8.1). However, Jores's opening speech notes the unexpected absence at the conference both of President Forsgren and founding member Kalmus, so their papers must either have been read by others or added to the proceedings without oral delivery. Moreover, of the twenty-one persons identified in the photograph, five are not named among the presenters, and fifteen authors are not identified by name in the photograph, leaving open the possibility that there were a number of attendees at the conference who were not presenting papers. There appear to be no women in the photograph, and Stoppel and Kleinhoonte, the two women who were earlier active at the Society's meetings and publishing, do not appear among the authors. This meeting seems to have been rather exclusively male and dominated by Germans.

Arthur Jores's opening speech begins with a greeting to all the participants, "especially our Swedish friends and our guest from Switzerland," which would include all the non-Germans presenting! He recounted for the audience that the *Societas* had been founded at Ronneby in 1937 with the broad purpose of studying plant, animal, and human rhythms and for this reason brought together botanists, zoologists, physicians, and meteorologists into one society.

---

[1] "Vorstand der Internationalen Gesell[s]chaft für biologische Rhythmusforschung 1949," in "Verhandlungen der dritten Konferenz der Internationalen Gesellschaft für Biologische Rhythmus-Forschung. Am 30. September und 1. Oktober 1949. Hamburg, Deutschland," ed. Hjalmar Holmgren, Jakob Möllerström, and Åke Swensson, *Acta Medica Scandinavica*, Suppl. 278 (1953): 7.

[2] However, Sollberger, *Biological Rhythm Research*, p. 316 lists it with the Society's other meetings, but as a "Scandinavian Conference" situated between the second and third meetings.

Figure 8.1 Participants in the third meeting of the *Societas pro studio rythmi biologici*, Hamburg 30 September–1 October 1949. Photograph published at the beginning of "Verhandlungen der dritten Konferenz der Internationalen Gesellschaft für Biologische Rhythmus-Forschung. Am 30. September und 1. Oktober 1949. Hamburg, Deutschland," ed. Hjalmar Holmgren, Jakob Möllerström, and Åke Swensson, *Acta Medica Scandinavica*, Suppl. 278 (1953).

Reproduced with permission of the current publisher of this journal, John Wiley & Sons, Ltd.

What was new at this 1949 meeting, he noted, was the presence of a mathematical approach, which promised to bring a greater exactness to the science of rhythms:

> Today I can tell you that the second day of our meeting will tell you something about the fact that indeed a mathematical formulation is also possible, and with that a step toward greater precision would be made. But the main problem, which is also the basic theme of most speeches at our conference, is still the question of the cause of the rhythmic swings of so many phenomena of life. . . . This topic, which already greatly occupied the meeting in Utrecht, will also be taken up at our meeting, since it is really one of the central problems. . . . [T]he questions of rhythm. . . . It is very striking that the Anglo-American literature has scarcely delivered any contribution to these problems thus far.[3]

We can see by this that the problem of working out the nature of the causation of biological rhythms, which Baranetsky and Pfeffer had raised around the turn of the century, and which Stålfelt had taken up in his doctoral dissertation (1921) and Holmgren had identified as a pressing problem in 1941, was still regarded at the 1949 meeting as the primary question to be addressed.

The proceedings of the Hamburg meeting were published in 1953, at which time the Society's status through 1952 was reported by the council, which then still formally comprised Forsgren, Möllerström, Jores, Gerritzen, and Holmgren, although in the meantime Hjalmar Holmgren had died in 1951, which was a blow to Swedish biological rhythms studies. Holmgren, scion of an academic family and an obvious protégé of Jakob Möllerström, was in many ways the young genius of the Karolinska Institute. A personal professorship in experimental histology was created for him there in 1947, with the expectation that he would help expand the histology department.[4] His death was not wholly unexpected, inasmuch as he showed signs of colon cancer already on a trip to Paris prior to the 1949 meeting in Hamburg, and the cancer rapidly metastasized. He was especially missed by Möllerström, who composed a memorial to him that was published with the conference proceedings in 1953, and

---

[3] A. Jores, "Eröffnungsaussprache," pp. 16–18 in "Verhandlung der dritten Konferenz," eds. Holmgren et al., p. 17: "Heute kann ich Ihnen sagen, dass der 2. Tag unserer Tagung Ihnen einiges darüber berichten wird, dass doch auch eine mathematische Formulierung möglich ist und damit ein Schritt zu grösserer Exaktheit getan würde. Aber das Hauptproblem, das auch das Grundthema der meisten Vorträge unserer Tagung darstellt, ist doch die Frage nach der Ursache des rhythmischen Schwingens so vieler Lebensphänomene. . . . Dieses Thema, das schon die Tagung in Utrecht sehr beschäftigt hat, wird auch unsere Tagung sehr in Anspruch nehmen, es ist auch wirklich eines der zentralen Probleme. . . . die Fragen des Rhythmus . . . Sehr auffallend ist es nur, dass das anglo-amerikanische Schrifttum noch kaum einen Beitrag zu diesen Problemen geliefert hat."

[4] An obituary for Hjalmar Holmgren appeared in the *British Medical Journal*, 7 April 1951, p. 765, noting his contribution to research on liver function, especially his collaboration with Wilander and Jorpes on finding the production of heparin in giant cells.

Arthur Jores spoke fondly of him when the Society convened the next meeting later in 1953, in Basel.[5] He was remembered as one of the vigorous biological rhythms researchers in Stockholm and an avid supporter of the *Societas*, which he had helped to found.

The council report noted that the war had made it impossible to convene a meeting of the Society between the Utrecht meeting in 1939 and 1949, acknowledging that a smaller conference was held in Stockholm in 1941, but that only Swedes had participated. It noted that during the war it was not possible to keep a list of members and that after the war problems with national currencies made it difficult to collect contributions from the majority of the members. The council then recognized a 5000 Swedish Crown donation by Director Martin Rind of Stocksund, which made possible the publication of the Hamburg meeting's papers in *Acta Medica Scandinavica*, and acknowledged the long interest in and support for biological rhythms research during the year by Axel Wenner-Gren, whose donations made it possible for the authors to obtain cost-free offprints of their contributions.[6] Thus, even though the Hamburg conference was dominated by Germans, the governing council was still composed largely of Swedes, and the publication of proceedings was funded by Swedes and appeared in the Stockholm journal *Acta Medica Scandinavica*, edited by Swedes Holmgren, Möllerström, and Åke Swensson, all of whom had presented at the meeting. A shift in leadership would begin with the next meeting in Basel.

When the *Societas* convened after the war, in Hamburg, it apparently intended also to resume the every-second-year cycle that was established in the first years, but the meeting planned for Basel 1951 was put off until 17–18 September 1953, when it was convened at the University of Basel by Arthur Jores and Felix Georgi, the local host. According to Jores, who gave the welcoming address, the two-year delay was occasioned by the unexpected death of Hjalmar Holmgren in 1951, whom Jores described as "our activist and valuable co-worker," bound in friendship to all the original members of the *Societas*.[7] If this is literally the case, then either Holmgren must have been shouldering a large share of the organizational effort of the Society's council, or else his death sufficiently disrupted planning efforts by his colleagues in Sweden, who still presided over the organization. The council list for 1953

---

[5] Jakob Möllerström, "Hjalmar Holmgren, 6/12 1905–21/2 1951," pp. 13–15 in "Verhandlung der dritten Konferenz," eds. Holmgren et al.; Arthur Jores, "Ansprache" pp. 18–20 in "Verhandlungen der vierten Konferenz der Internationalen Gesellschaft für Biologische Rhythmus-Forschung. 18.–19. September 1953. Basel, Schweiz" eds. W. Menzel, Jakob Möllerström, and Ture Petrén, *Acta Medica Scandinavica*, Suppl. 307 (1955).

[6] "Vorstandsbericht für die Jahre 1939–1952," pp. 9–10 in "Verhandlung der dritten Konferenz," eds. Holmgren et al.

[7] Jores, "Ansprache," p. 19.

still shows Forsgren as president and Möllerström as vice president, but now the other council members were all non-Swedes: Jores, Gerritzen, and Georgi, with Kleinhoonte and Kalmus as alternates, and Werner Menzel replacing Holmgren as secretary and treasurer. This cohort convened the council at Basel, and its minutes witness the transition in leadership from Swedish domination to a more international mix. After Forsgren paid his memorial tribute to their former colleague and council member, he declared that he was stepping down as president and being replaced by Jores, who was voted in as president elect. Georgi replaced Möllerström as vice president, and the next meeting was set for Stockholm, with Ture Petrén as convenor and local host. The possibility of publishing the proceedings of the 1953 meeting with Karger in Basel was discussed, but in the end Möllerström was asked to inquire about continuing to publish them in Stockholm, and they came out as another supplement to *Acta Medica Scandinavica*.[8] This time, however, the German Menzel joined the Swedes Möllerström and Petrén as editors, another sign that the Society's orientation was becoming more international, even if it was still publishing in Stockholm and perhaps still depending on Axel Wenner-Gren's support. The brief treasurer's report that was published does not specify where the Society's money came from, but Möllerström's recommendation at the council meeting to make Wenner-Gren an honorary member suggests his continued involvement.[9]

The final act of the council as recorded in the published minutes is a telling indication of the Society's true internationalization. A committee was formed "to address the question of nomenclature" comprising some members who had attended the previous meeting (Forsgren, Jores, Kalmus, Menzel, Georgi, and Wachholder), but also some new names (Aschoff, Bartels, Bünning, Halberg, and Kleitman). Of these last, Nathaniel Kleitman was a veteran American biological rhythms researcher, best known perhaps for his research on sleep, some of which he had conducted in Kentucky's Mammoth Cave, and Bünning was a seasoned German researcher, who for whatever reason had not previously taken part in the *Societas*.[10] The German Jürgen Aschoff and the recently naturalized American Franz Halberg were newcomers, but destined to become salient voices in the field in the next several decades. Halberg, in particular,

[8] "Vorstand der Internationalen Gesellschaft für Biologische Rhythmusforschung 1953," p. 9 in "Verhandlungen der vierten Konferenz," eds. Menzel et al., Swedes Arborelius and Nilsson are still listed as auditors.

[9] "Vorstandssitzung der Internationalen Gesellschaft für Biologische Rhythmusforschung," pp. 11–12 in "Verhandlungen der vierten Konferenz," eds. Menzel et al., p. 12: "Auf Antrag von Möllerström wird beschlossen, Herrn Dr. h. c. Wenner-Gren die Ehrenmitgliedschaft zu verleihen." [At the request of Möllerström it was decided to bestow an honorary membership on Dr. h. c. Wenner-Gren.]

[10] Ibid., p. 12. Nathaniel Kleitman, "Biological Rhythms and Cycles," *Physiological Reviews* 29 (1949):1–30, aptly illustrates his grasp of the work that had been done in the field, including the American contributions with which Jores and others seem to have been less familiar.

used his membership on this committee to create a nomenclature that he would wield to transform the Society's focus from "biological rhythm research" to *chronobiology*.[11]

The attendance at the 1953 meeting also signaled a shift in the Society's composition and its growth in the postwar years. Whereas the conference photograph in Hamburg showed thirty-one heads, its counterpart four years later shows sixty-nine, at least six of whom are women, and they do not appear to be the old guard (Figure 8.2). The conference program in the published proceedings does not include Rose Stoppel and Antonia Kleinhoonte, and the only Swedes I recognize in the list are Möllerström, Swensson, Petrén, and his protégé Arne Sollberger. The latter two would continue to play an active role in the *Societas*, but clearly its composition and leadership were no longer dominated by Swedes. Instead, one sees a number of new names, representing more nationalities than were evident at the early meetings.

Felix Georgi's opening speech at the conference recognized the internationalizing trend of the *Societas* and perhaps also the broadening of the field of research. He began by noting that this was the first time the group had met in Switzerland and acknowledged both the presence of President Forsgren, one of the Society's founders, and the absence of "another of the Swedish pioneers, Hjalmar Holmgren, no longer with us."[12] Georgi pointed out that the foundation meeting had counted some twenty "rhythm enthusiasts," to use Forsgren's term, but that interest in "our problem" had grown considerably since 1937, and now he welcomed participants from both sides of the Atlantic Ocean, slipping easily from German to English and French:

> We are welcoming very heartily our guests from the United States and from Great Britain. Wir begrüssen herzlich die Gäste aus Norwegen und Finland und vom Libanon . . . De même nous sommes très heureux pouvoir sluer les Confrères venant de la France . . . Wir begrüssen herzlich unser Gäste aus den Niederland . . . weiter unsere Gäste aus Österreich und das stattliche Kontingent von Teilnehmern aus Deutschland. . . . Last but not least aber möchte ich auch die Schweitzer Kollegen begrüssen.[13]

The fifth meeting of the *Societas* was convened in Stockholm in 1955, organized by Petrén and Sollberger, and they edited the proceedings, which were

---

[11] "Glossary of Chronobiology," eds. Franz Halberg, Franca Carandente, Germaine Cornelissen, and George S. Katinas, *Chronobiologia* 4, Supplement 1 (1977). It is clear from the fact that Halberg is listed as first editor, that staff at the University of Minnesota are listed prominently in the acknowledgments, and that Franz Halberg holds the copyright that he had come to monopolize the work of the International Committee for Nomenclature to which he was appointed a junior member in Basel in 1953. As Cambrosio and Keating, "The Disciplinary Stake: The Case of Chronobiology," p. 342, point out, Halberg's attempt to control the language can be seen as a normative measure to direct the development of the field.

[12] Felix Georgi, "Eröffnungsansprache," pp. 15–17 in Verhandlungen der vierten Konferenz," eds. Menzel et al., p. 15: "ein anderer der schwedischen Pioniere, Hjalmar Holmgren, nicht mehr unter us."

[13] Ibid., pp. 15–16.

Figure 8.2 Participants in the fourth meeting of the *Societas pro studio rythmi biologici*, Basel 18–19 September 1953. Photograph published at the beginning of "Verhandlungen der vierten Konferenz der Internationalen Gesellschaft für Biologische Rhythmus-Forschung. 18.–19. September 1953. Basel, Schweiz," ed. W. Menzel, Jakob Möllerström, and Ture Petrén, *Acta Medica Scandinavica*, Suppl. 307 (1955). Participants were not identified in the photograph when it was published, but the second man to the right of the left pillar, in the back row, has been identified as Alexander Pierach.

Reproduced with permission of the current publisher of this journal, John Wiley & Sons, Ltd.

eventually published in Stockholm. Even so, Swedish leadership in the field was yielding to German, British, and increasingly American researchers.[14] Möllerström retired from his academic position in 1957, leaving Petrén to carry the torch. His protégé Sollberger failed to attain the doctoral degree in Stockholm and moved to the United States to continue his research, where he remained the Society's treasurer, at least until 1965 (when he identified himself as such on the title page of his monograph *Biological Rhythm Research*) and became involved with the *Journal for Interdisciplinary Cycle Research* as one of the chief editors 1978–88. Swedes stayed active in biological rhythms research, particularly in photobiology and especially in the clinical and industrial–medical aspects, but now also in molecular biological research, which largely took over from the behaviorists the search for endogenous clock mechanisms. Notably, the Swede Ragnar Granit was awarded the Nobel Prize for Physiology or Medicine in 1967 for his research on neural response to light stimulus, color, and frequency. More central to chronobiology, his compatriot Lennart Wetterberg undertook careful study of the role of melatonin in animal rhythms and, though now retired from the Karolinska Institute, is still an active participant in the field. But after the 1955 meeting, organizational aspects of the Society were no longer closely associated with Sweden, and a list of related meetings that were held during the 1950s and early 1960s shows both the general increase of activity internationally in rhythms-related research and scholarship and the greater specialization that was taking place: The sixth meeting of the *Societas* was held in Austria (1957), the seventh in Italy (1960), and the eighth again in Hamburg (1963); the first International Photobiology Congress was held in Holland in 1954 and subsequent ones in Italy (1957) and Denmark (1960); the International Bioclimatology Conference was first held in Austria in 1957 and subsequently in England (1960) and in France (1963); in the United States, biological rhythms researchers convened a Symposium on Application of Autocorrelation at Woods Hole in 1949, a Symposium on Cycles in Animal Populations in 1954, a Symposium on Biological Chronometry in 1956, a Symposium on Biological Rhythms in 1957, and a Conference on Photoperiodism in 1957. The 25th Cold Spring Harbor Symposium on Quantitative Biology in 1960 was devoted to the subject "Biological Clocks." The symposium brought together leading researchers from many countries to present the state of rhythms research in a wide variety of aspects representing a multiplicity of methods, models, and applications. Key philosophical and

[14] *Society for Biological Rhythm. Reports from the 5th International Conference, Stockholm Sept. 15 – 17th 1955*, ed. A. Sollberger, T. Petren (Stockholm: no publisher, 1961).

methodological issues were presented, and consequently the meeting is generally regarded as pivotal for the development of biological rhythms study as a discipline.[15]

Following the Cold Spring Harbor Symposium, the center of gravity of biological rhythms studies shifted to the Anglo-American world. When the International Society for the Study of Biological Rhythm finally convened in the United States (1971), in Little Rock, Arkansas, its president, Franz Halberg, formalized a movement to rebrand biological rhythm studies as *chronobiology* by formally changing the name of the *Societas pro Studio rythmi biologici* to the International Society for Chronobiology and by promoting the creation of the *International Journal of Chronobiology* with Alain Reinberg, Hugh Simpson, and himself as co-editors (1973) and a second journal, *Chronobiologia* (1974).[16] The Society's current journal is *Chronobiology International*, founded in 1984, also during Halberg's presidency.[17]

---

[15] Sollberger, *Biological Rhythm Research*, pp. 316–20 gives a list of meetings and journals related to biological rhythm studies.

[16] Cambrosio and Keating, "The Disciplinary Stake: The Case of Chronobiology," pp. 336–40. The term *chronobiology* technically includes biological rhythms research in a slightly larger conception of a discipline or research field that has as its subject the temporal nature of living things. Besides cyclical phenomena this includes such linear temporal development as maturation and aging and, in principle, also evolutionary development. However, its core identity remains rhythmic phenomena and their applications, so there is little advantage, in my view, to maintaining a formal distinction between chronobiology and biological rhythms research. This distinction is suggested by the names of leading journals, which reflect the earlier struggles within the field.

[17] Michael Smolensky, "Editorial," *Chronobiology International* 1 (1984): i.

9

Conclusion

The formal establishment of an international academic society to facilitate research in a common scientific field marks a significant stage in the recognition of that field. The foundation of the *Societas pro Studio rythmi biologici*—an international society devoted specifically to fostering a multidisciplinary approach to the study of the rhythmic behaviors of biological organisms—signals the beginning of the formation of a new disciplinary entity, known today as *chronobiology*. Prior to the founding of this Society, researchers in many countries worked within their own disciplines of botany or zoology or electrical engineering or in defined subdisciplines of these, such as histology and endocrinology, often neglecting the related work of colleagues in other fields. The founding members of the *Societas* recognized this problem from the beginning and sought to create both a common forum for sharing the results of their work across traditional disciplinary boundaries and a common outlet for publication of the proceedings of their biannual meetings.

That this important, early step in creating a disciplinary identity took place in Sweden in 1937 can be explained by a number of historical, economic, and institutional circumstances: the presence of a core group of biological rhythms researchers in Swedish academic and medical institutions; a historical tradition of interest in the physics of helio-atmospheric phenomena and the rhythms of light and darkness within Swedish science and medicine; and a relatively pleasant and accessible, economically strong, and politically stabile climate for hosting the formation of a new academic society. These factors supported the initial meeting of the charter members at Ronneby Brunn and also acted to support the new Society during the difficult years immediately before, during, and after the Second World War. This explains why Swedish scientists played such a prominent role in the initial steps to create a new international discipline and continued to support it economically and to dominate its governance until the 1950s, when the *Societas* emerged as an international academic society in fact as well as in name.

Index

— Index —

WITHDRAWN

DATE DUE

GAYLORD			PRINTED IN U.S.A.